Women Astronauts

by

Laura S. Woodmansee

To Paul: You make my dreams come true.

We acknowledge the financial support of the Government of Canada through the
Book Publishing Industry Development Program for our publishing activities.
Published by Apogee Books an imprint of Collector's Guide Publishing Inc., Box 62034, Burlington, Ontario, Canada, L7R 4K2
Printed and bound in Canada
Women Astronauts
by Laura S. Woodmansee
ISBN 1-896522-87-4

Women Astronauts

An Apogee Books Publication

Table of Contents

Acknowledgements

I have so many great people to thank for helping and encouraging me while I wrote this book. Thank you to my husband Paul, Mom and Dad, and my sister Tracy for their support and enthusiasm about this book, and my career. Thank you to Richard Godwin and Robert Godwin of Apogee Books for giving me the opportunity to write this book.

Thanks to my friends, Rachel Zimmerman Brachman and her husband Scott Brachman, and Gindi French for being so excited about this book. And a special thanks to Celeste Chan Wolfe and her husband Robert Wolfe for the most important introduction of my life.

There are so many people who helped me research this book. Thank you to Benny Cheney of the NASA/JSC Media Resource Center for all the great space videos. To Reference Librarian D. Chris Cottrill of the National Air and Space Museum for finding great old magazine articles for me. To NASA/JSC Media Resource Center's Jenny Ann Crawford, who searched extensively and found several hard-to-find images for me. To Dr. Deborah L. Harm of NASA/JSC Neuroscience Laboratory who co-wrote an article on space medicine that got me started on the topic. To NASA/JSC Newsroom Manager Eileen Hawley for setting up interviews with astronauts and answering all sorts of astronaut questions. To Yuri's Night co-founder Loretta Hidalgo for being encouraging and introducing me to Apogee Books.

To Richard Jennings, director of the UTMB Aviation Medicine Center for sending me his wonderful space medicine article. To Dayna Steele Justiz, President of the Space Store, for a great interview and her excitement about this book. To NASA/JSC Newsroom Coordinator Kacy C. Kossum for setting up interviews, finding video and audiotapes, and for her support. To NASA Chief Historian Roger D. Launius for getting me started with the FLATS issue. To the United Space Alliance/Astronaut Appearance Office's Lucy Lytwynsky who arranged astronaut interviews for me and answered all kinds of questions about the astronauts. To Simon Mansfield at SpaceDaily.com for publishing my work. To Dotti Martin of the Girl Scouts USA for sending me information about the Aerospace badge and my very own badge! To Bear Ride and Terry McEntee for arranging a great interview with Sally Ride.

To National Air and Space Museum Archives Deputy Reference Chief Brian Nicklas for sending me tons of articles about NASA's first women astronauts. To NASA Chief Archivist Jane H. Odom for finding all sorts of women astronaut information sources for me. To Margaret Persinger at NASA/KSC's Still/Video Public Affairs office for finding images for me. To Women of Space CEO Alisa Schreibman McKnight for her introduction to Robert and Richard Godwin. And thank you to Margaret A. Weitekamp, Assistant Professor at Hobart and William Smith Colleges Women's Studies Program for opening the world's eyes to the real FLATS story. A special thank you to Astronauts Claudie André-Deshays Haigneré, Ellen Baker, Yvonne Cagle, Jan Davis, Bonnie Dunbar, Anna Fisher, Eileen Collins, Linda Godwin, Ellen Ochoa, Kalpana Chawla, kindred spirit Millie Hughes-Fulford, Janet Kavandi, Wendy Lawrence, Shannon Lucid, Heidi Stefanyshyn-Piper, Sally Ride and Janice Voss. And thank you to my USC Annenberg School of Journalism Graduate School professors for helping me to become a better writer, especially to Frank Mottek and Cinny Kennard who encouraged my interest in writing about space, no matter what.

Preface

Why I Wrote This Book

I have this recurring daydream that I'm floating above the Earth in a space suit. I can hear myself breathing as I look up at the stars and down at the Earth below my feet. I've always wanted to be an astronaut.

I wrote this book because it has always been my dream to be involved in space exploration. All my life, I've loved the idea of seeing the Earth from space and traveling to other worlds. I've always loved space, and I always will.

When I was 10, my grandfather gave me a book about the planets of our solar system. This was before we knew for sure that planets outside our solar system existed. He wrote a note inside the cover, "Enjoy this book, and learn all about our universe. Look at the pictures and imagine taking a trip to the planets." When I read the book, it made me want to learn more. Grandpa always encouraged my interest in space and science. He is one of the main reasons that I'm a science writer today.

That book about the universe has sat next to me on my desk while I wrote *Women Astronauts*. It's a symbol that my grandfather is still with me, encouraging me. In the same way, I hope that this book deepens your interest in space exploration and encourages you to pursue your dreams.

I began my research for *Women Astronauts* by looking for other books about women who have flown in space. I was surprised and disappointed to find that there are very few. The books that I found were very good, however, each of them focused on only one woman astronaut, or a small group of astronauts. There were books on Sally Ride, Shannon Lucid, and Ellen Ochoa, but none covered *all* the women of space.

It dawned on me that I had a great opportunity here, a chance to do something different. There has never before been a book like the one that you are holding in your hands! A book that covers *every* woman who has ever flown into space, and some that will fly soon. This book is a history of women astronauts, a guide to current ones, and hopefully an inspiration for future astronauts. I hope that reading *Women Astronauts* inspires you the way writing it has inspired me.

There is a reason that astronauts are the superstars of space exploration. They are the pioneers lighting the way for the rest of us.

Introduction

The word *astronaut* comes from Greek words meaning "star sailor." Astronauts are extraordinary people, but they are not gods. They are people just like you and me who worked hard to follow their dreams. Each astronaut has traveled a different path to get to where they are now. Their stories are fascinating and inspiring.

Women astronauts have a special place in the history of space exploration. Their contributions need to be recognized. This book aims to do just that, and to inspire girls, young women, and everyone else to explore space.

Why explore space? Because it's there and human beings have always wanted to know what's over that next hill in the distance. We have always adapted to different environments because it's what we do best. We are explorers making our first steps into the universe. Astronauts, both women and men, are our ambassadors to space.

We watch the ceremony of launch on television and we hear the words, but nothing can compare to the reality of being strapped into a chair and launched through the atmosphere. What's that like? The sounds? The shaking? Weightlessness?

This book begins with a space shuttle launch and with first hand accounts from women astronauts about what it's like to be in space. I begin with this because it's what I find most fascinating about these women. They've experienced something that few of us have.

To understand where we are now, we need to understand where we've been. Chapter 2 tells the story of a group of women who had the right stuff to become astronauts, but the wrong gender. We need to look at the gender bias of the past in order to appreciate how much better things are today.

Chapter 3 is about the Space Age, and the Space Race, and how the first women astronauts made history, as well as some of the more recent achievements women astronauts have made.

Meet the women astronauts in chapter 4. Read about their lives and what sparked their interest in space and how they became astronauts. Space flight might be the most exciting and fun part of the job, but there's much more to each of these women than their missions into space. Those I've talked with feel deeply that it is an honor and a privilege to serve as astronauts. By risking everything, they are giving us the gift of advancing the human exploration of space.

In chapter 5, learn about medical issues that may affect women space travelers now and in the future. This stuff hasn't been talked about in the popular press.

Finally, in the last chapter learn how you can get involved today and what's in store for you if you become an astronaut. You can be involved in space exploration in many ways and this chapter will give you some suggestions how and where to continue your adventure.

Astronauts don't become astronauts just to be in space. That's a small part of the job for most of them. Most of their time is spent on the ground working on experiments, software, and hardware. The job is really about learning and exploring and making sure that flight crews will be safe in space and able to do their job of bringing knowledge to the world.

The late astronomer Carl Sagan called Earth, "the shore of the cosmic ocean." Our "Spaceship Earth" keeps us safe as we travel through space, around the sun and around the galactic core, and into the universe. In this way, we are all space travelers. So, consider yourself an astronaut already.

My wish is that you enjoy this book and that all your space travel dreams come true. You're never too young or too old to get involved in something that you love. So reach for the stars.

1

Living in Space

Lift Off

Seven-and-a-half minutes before launch, the tower rolls away from the Shuttle. At T minus 6.6 seconds, the Shuttle's three main engines start. At this point, there is no looking back ...3 - 2 - 1 ...Lift-Off!

The Space Shuttle *Columbia* blasts off on mission STS-1

You see the bright flash of light clearly as the engines fire. It's eerily quiet as the sound travels the three miles to reach you. Then when it does, you feel the thunderous roar in your chest and in your feet as well as in your ears. On board, however, the sound is immediately obvious, but it's not much louder than riding inside a noisy airplane.

As the shuttle rises into the sky it leaves a bright contrail behind it. Eight seconds into the launch the shuttle rolls 120 degrees to the right. If you have a telescope or binoculars, about 50 seconds into the flight you can see a thin cloud of water vapor form near the nose as the shuttle breaks the sound barrier.

The experience is beautiful to watch. Onboard, the astronauts get a shaky ride while the solid rocket boosters are burning. The crew is pushed back into their seats by an acceleration of three times the force of gravity (G's). After two minutes of flight the boosters separate and fall back into the sea for recovery and re-use. From then on, the ride is smoother as the astronauts are propelled into space on the three space shuttle main engines.

Eight minutes into flight, the orange external tank separates, and the shuttle orbiter is left to fly alone into space. Thirty seconds later the main engines are cut off, and the shuttle is finally floating in orbit at the end of another successful launch.

When the engines stop burning, the astronauts experience weightlessness. It's an abrupt change from 3G's to weightlessness that Astronaut Story Musgrave once called "magical."

It's taken a lot of work to get into space. Before we take a look at what life is like in space, let's take a brief look at how our astronauts get there.

Before a flight, astronauts work on several different jobs helping to get hardware, science experiments, software, other astronauts, and themselves, ready for a mission.

The Crew of STS-76 beside the Space Shuttle *Atlantis* at a dress rehearsal for launch

Getting into orbit is a challenge, but putting people in space is even more difficult. Launch is the first step into space. Some American, Russian and European astronauts fly on the Russian *Soyuz* launch vehicle. Today, nearly all astronauts are headed to the International Space Station (ISS) and the space shuttle is how most astronauts from Earth get there.

The Space Shuttle

Enterprise, NASA's first space shuttle, was built in the 1970's as a test vehicle and used for launch pad and landing tests at Dryden Flight Research Center at Edwards Air Force Base in California. Named after the fictional "Star Trek" starship, NASA's Space Shuttle *Enterprise* was never intended to fly into space.

After the *Enterprise* was built and tested, improvements were made and engineers built the *Columbia*, then *Challenger*, *Discovery*, *Atlantis*, and finally the *Endeavour*. As of this date, there are no plans to build a seventh space shuttle. But who knows, maybe the flagship of NASA's next generation of space ships will be called the *Enterprise*.

The Space Shuttle *Enterprise* with the crew of Star Trek

The three space shuttle main engines work together to deliver up to 375,000 pounds of thrust each (170,097 kilograms) at sea level. (A total of 1,125,000 lbs) The engines use so much fuel that the shuttle itself can't carry it all. So, attached to the belly of the shuttle is a large orange external fuel tank that supplies the shuttle's main engines with liquid hydrogen and oxygen during the launch.

Even with all of this power, the shuttle needs an extra boost to get into space. That's why there are two solid rocket boosters strapped on to the external fuel tank. The two boosters combined provide 5,300,000 pounds of thrust (2,387,000 kilograms) for the first two minutes of launch. At that point, they separate from the external fuel tank and fall into the ocean. The used boosters are later hauled to shore and fixed up for the next shuttle launch.

Together, the space shuttle main engines, the external tank, and the two solid rocket boosters launch the Shuttle into space. Total thrust of main engines plus boosters at launch is 6,425,000 lbs.

The launch uses up so much energy in so short of a time, it's like surfing on the front of a barely controlled explosion all the way into space. It's exciting, but scary. That's why everyone involved in launching people into space is so careful. Launch controllers check every piece of equipment again and again to make sure the spacecraft is safe for the crew to ride into space.

Launch is the most dangerous part of the whole mission. Astronaut Anna Fisher says, "The really scary part is the first two minutes until the solid rocket boosters come off, because there really isn't a lot you can do. After that you have quite a few options depending on what the situation is. Once you have main engine cut off there are very few things at that point that are going to kill you. It would have to be something really major."

Making a flight safe for people is also very expensive. NASA calls this making it "man-rated." Thousands of hours of testing, analysis, and double-checking are done before anything or anyone gets to fly on the Shuttle. The cost is high, but it would be even higher if a device malfunctioned during launch or in flight. A glitch, failure, or error could get someone killed.

People need lots of air to breathe, water to drink, food to eat, power to keep warm, and room to move around. Besides, people damage easier than satellite hardware. All of this stuff costs money to lift into space, about $10,000 a pound. So, a half-pound serving of shrimp cocktail costs $5000 to serve in orbit. That's an expensive meal by any standard. Whether it's a gourmet meal or cheap kibble, it will still cost the same in orbit.

Countdown

The launch process is a ceremony that must be nearly perfect every time. Astronauts are our ambassadors to space, and launch is their ceremonial send off. There is zero tolerance for failure. Everything is planned out to the millisecond. And if everything doesn't go as planned, the ceremony is called to a halt.

The Launch Countdown Clock with the Shuttle on the pad

"Cape Crusaders," also known as Astronaut Support Persons (ASP's), spend about four days at the Kennedy Space Center as part of the launch team helping to prepare a Shuttle crew for launch. Cape Crusaders are astronauts and know the space shuttle systems inside and out. They help launch control make sure that all systems are working correctly.

"The Cape Crusaders are actually making sure the shuttle is put together right," says Payload Specialist Millie Hughes-Fulford. "Acting with the workers. Keeping morale up. They are kind of like the astronaut representatives for the astronauts. And that's key. You need someone representing you."

During launch and re-entry into the Earth's atmosphere, astronauts wear the Advanced Crew Escape Suit (ACES) and the Crew Altitude Protection System (CAPS) which includes a bright orange suit that acts as a pressure suit, an anti-exposure suit, and a g-suit to prevent blackouts while experiencing high acceleration rates also known as "G's." Astronauts also wear gloves, a "snoopy cap" for communications, a helmet and boots.

Janet Kavandi in her "Snoopy" communications cap

In case the astronauts need to evacuate the space shuttle orbiter at a high altitude during a takeoff or landing emergency, they wear escape gear over the CAPS. If astronauts were to bail out of the ship after launch, they would probably end up in the ocean and the ACES would keep them warm in the cold water. Survival gear includes emergency oxygen, a parachute, a life raft, flotation devices and two liters of drinking water. Astronauts also wear a survival vest that holds a radio/beacon, a knife, a flare kit, a smoke flare, sea dye marker, and a signal mirror.

Launch day begins early for everyone involved. Crewmembers are awakened 5 hours before launch. After a shower and breakfast, the astronauts suit up for lift-off. Before they board the "Astro-Van" and head out to the launch pad, the crew waves to the news media and others gathered to take pictures and wish them good luck.

STS-102 crew at the AstroVan (Susan Helms in the middle)

Once they get to the pad, it's practically empty. Astronauts arrive at the launch pad two and a half hours before take-off and ride an elevator 200 feet up the launch tower to the "White Room." There is no one on the pad except the crew, the Cape Crusaders, and a few technicians who strap and seal the astronauts into the Shuttle and make the final safety checks. These dozen or so people are the only human beings for miles. It's just them and the rocket loaded with explosive liquid hydrogen and liquid oxygen.

The Russians have a different philosophy for their launches. Hundreds of people go with the astronauts to the launch site and say goodbye and good luck as the astronauts go up the gantry and into the capsule. Then the pad is evacuated.

The astronauts put on their parachutes and communications gear and crawl through the space shuttle's hatch. Since the Shuttle is vertical for launch, it's difficult to get into the seats, especially wearing the puffy ACES suit. So the Cape Crusaders and suit technicians help the astronauts get into their seats and safely buckled in for launch.

An hour before launch, technicians seal the hatch and check the Shuttle for air leaks. When they are satisfied that everything is safe, the launch team travels far away from the pad. During launch, no one is allowed within three miles of the space shuttle in case of an emergency.

Countdown is never routine. Everyone involved wonders if they've done enough. Minor glitches can hold a countdown until fixed. Major problems will stop the count, and they will have to try again another day. Aborted launch attempts are common, and bad weather at the launch site or one of several emergency-landing sites, is the most common reason to stop a launch.

Once the crew gets into space, they open the payload doors to radiate all the excess heat gained during launch. Next, they turn on the life support systems and check to make sure everything is working well. They stow seats and hardware used for launch and re-entry. Before they get to work, however, everyone takes a look out the window.

The View

One of the benefits of being in space is the wonderful view of the Earth from orbit. Anna Fisher says that the best part of being in space was, "just floating and looking at the Earth, it's just so incredibly beautiful."

Susan Helms looking at Earth through a window on the ISS

Many astronauts spend what little free time they have staring out the window, especially on their first flight into space. America's first woman in space Sally Ride says that, "In space you get a spectacular view of Earth and it's a constantly changing view and it's a view that you simply can't get from anywhere else. Astronauts would love to spend all of their time with their eyes glued to the window."

After her first mission, Ride told reporters, "You spend a year training just which dials to look at. And when the time comes, all you want to do is look out the window. It's so beautiful."

Astronaut Shannon Lucid spent 188-days on the *Mir* space station. To get a good look at the Earth, she had to make a special trip to another module of the station, "Often it seemed like you would go to the window to look out and find out, 'Oh we're over the ocean again.' I never realized, I mean you know, but it doesn't really hit home, how much of the Earth is really water. I feel like any time you had a moment to look out, there you were over the ocean once again. It was always neat to have a really great window that you could float over to and look out and see the Earth."

Oblique view of the Galapagos Islands showing the atmosphere of Earth

"I have the most spectacular view of the world right in front of my eyes," says Astronaut Wendy Lawrence of her Shuttle flights. "So when I want to relax, I'm going to put my nose to the glass and look out the window."

"I think it's given me a different perspective on how I view our world," says Astronaut Jan Davis. "It made me realize it was one home, one planet for everyone. It also made me realize how fragile the environment is. You see how thin the atmosphere really is and what we're doing to pollute it, and that just makes you more sensitive to it. So it just gave me a whole other perspective on where we live."

"It looked alive," says Millie Hughes-Fulford. "And I finally figured out it was because it looked like it had a glow to it. The same glow that a living cell has."

Looking at the thousands of photographs and videos the astronauts have made of their trips into space, we can easily imagine ourselves watching the Earth from orbit. Several IMAX movies have been made for just this reason. *Blue Planet* and *The Dream Is Alive* focus on the beauty of the Earth from space and give you a sense of what it's like to be an astronaut in weightlessness.

Zero G

Once the Shuttle's main engines stop firing, the astronauts experience zero gravity, also called weightlessness. On Earth, you experience something similar when floating in the water. Your buoyancy counteracts gravity in the water. Weightlessness is a little like being in the water, only more extreme.

Like a swimmer, when you are weightless in space, any slight push can send you moving in one direction or another. But in space, there isn't the resistance of water to hold you in place. The air in the cabin will resist your motion a little bit, but not nearly as much as water would. It takes astronauts a few hours to learn how to move in weightlessness.

Susan Helms lifts a treadmill on the ISS

Wendy Lawrence loved, "being able to fly around like a bird. It's just very fun to not feel the effects of gravity, to just, no kidding, be able to move yourself out to the middle of the cabin and let go and completely relax and be perfectly suspended right there in the middle of the air. It's just fun."

Zero gravity can be disorienting and takes a little time to get used to. Astronauts use a mix of handholds and floating to move around inside the cabin. What surprises some astronauts is the rotational motion they get when pushing off. People don't push off perfectly straight, so each time they get a little spin. Some astronaut's feel like everything is upside down and they are hanging in space. But, once they get used to it, weightlessness is fun.

You can spin in place, and perform feats that would be amazing here on Earth. On almost every mission, astronauts have made videos of themselves floating in midair, spinning in place, and making their food, pens, and anything else they can think of into zero gravity toys.

Marsha Ivins and Robert Curbeam floating on the Shuttle mid deck

"Astronauts tend to have a really good time in weightlessness and that was true during both of my flights," says Sally Ride. "During the off hours we'd basically have a good time doing somersaults in the middle of the cabin or floating peanuts from one side of the room to the other."

When in orbit, the shuttle and everyone inside are actually falling towards Earth. The ship's forward momentum, however, is keeping it in orbit and weightless. This freefall is a confusing feeling to the human body and can lead to space sickness.

Space Sickness

In space, things are not always perfect. About half of the astronauts feel sick when they first arrive in orbit. Space adaptation sickness is similar in women and men.

Unfortunately, there is no way to tell which astronauts will be affected. "I vomited my guts out," says Millie Hughes-Fulford of her 1991 Space Shuttle Flight. As a payload specialist, Hughes-Fulford had gone through nine years of rigorous flight training with her crewmembers. "I didn't get sick the whole time before." A shot of anti-vomiting medication helped take away the nausea. Her advice to new astronauts is, "Take the shot."

Anna Fisher also had space adaptation sickness the first few days of her first mission, "I remember thinking that I don't ever want to come back." Thankfully, medicine helped in her case too. "By the fourth day it was 100% gone and it was so much fun, and I remember wondering why I felt that way. But then, I used to get seasick when I went scuba diving, so I learned to just deal with it."

Astronaut and medical doctor Ellen Baker says that some of the astronauts "feel a little queasy and maybe even vomit the first two to three days of the mission. But that generally passes and we actually have some very good medicine for them. So, it's not that much of a factor any more."

Typically, by the third day of the mission everyone has recovered from space sickness. Medicines help, but they don't take away all the symptoms, so researchers are working on ways to stop it before it starts. There is more than one reason why astronauts get space sick.

The first cause is the vestibular system. The fluid inside your inner ear that tells you which way is down, and which way is up. Without gravity there is no "down" and the inner ear, which controls a person's balance, becomes confused. Some people get vertigo when this happens. One astronaut reported that when he opened his eyes after closing them for a while, the cabin seemed like it was spinning for a few seconds.

Have you ever "fallen" on a roller coaster ride? Did your stomach feel like it was rising up? That's what freefall feels like. There is no up or down in space, and you are constantly "falling" towards Earth. Only your spaceship's forward momentum keeps you in a circular orbit instead of falling back into the Earth's atmosphere. It takes astronauts some time to get used to the feeling that they are constantly falling.

The feeling of vertigo and falling combined with confused nerves can make an astronaut feel space sick. The causes and symptoms are similar to carsickness, airsickness, or seasickness.

The body's proprioceptive system can be fooled in the weightless environment of space. The proprioceptive system is made up of the nerves in the body's joints and muscles that tell the brain where our arms and legs are without looking at them. "The first night in space when I was drifting off to sleep," recalled one *Apollo* astronaut, "I suddenly realized that I had lost track of my arms and legs. For all my mind could tell, my limbs were not there. However, with a conscious command for an arm or leg to move, it instantly reappeared, only to disappear again when I relaxed."

When an astronaut first gets into orbit, her head feels stuffy. On the ground, the body relies on gravity to keep fluids from the torso and legs away from the head. When gravity disappears suddenly, the head can get an excess of these fluids. This can lead to light-headedness, clogged sinuses, a stuffy nose, and can feel like a bad head cold for the first day or two.

The worst part of being in space for Wendy Lawrence was, "The adjustment from a gravity environment to an environment where you don't feel the effects of gravity. Your body has to go through an adjustment period. So for me that's about 24 hours, and my inner ear doesn't like that the gravity vector has been removed. So you just have to be kind of careful when you go through that adjustment phase. The fluid has shifted to your head, you have a headache, your nose is stuffed up, and your inner ear is confused. Once you get past that period, it's very comfortable to be in space. That's a small price to pay."

Space sickness is unpredictable. An astronaut who is fine in an airplane may be all right on one space mission and space sick on the next.

Luckily, space sickness is temporary and astronauts recover from it quickly enough to do their work. Even so, it's important to NASA to put an end to space sickness. In addition to making astronauts feel lousy, it could put a mission at risk. In 1997, the National Space Biomedical Research Institute (NSBRI) was established to study space sickness and how the body adapts to space. NSBRI hopes to find ways to make the transition into space easier.

Space Wardrobe

NASA has created a whole line of comfy clothing for astronauts to wear in the Shuttle and on the International Space Station (ISS). After the astronauts reach orbit, they change out of their bright orange ACES suits and into something more comfortable. From launch suits to spacesuits, astronauts have all sorts of clothing to protect and comfort them for every occasion.

NASA describes the Shuttle as a "shirt-sleeve" environment. For Shuttle and ISS duty, astronauts wear IVA (intravehicular activity) clothing such as long and short-sleeved T-shirts, pants, shorts, jackets, flight suits, and soft slipper socks. In general, pretty comfy stuff.

Sometimes a crew will wear matching clothing, usually for a celebration or media event.

The STS-99 crew in a starburst pose

Spacewalking

Only a few astronauts have been lucky enough to put on a space suit and take a "walk" in space. In July 1984, Soviet Cosmonaut Svetlana Savitskaya took her first "steps" in space and became the first woman to walk in space. U.S. astronaut Kathryn Sullivan followed her in October of that same year with a three-and-a-half hour spacewalk.

Why spacewalk? Because things need to be built, fixed and maintained in the vacuum of space. Sometimes satellites are grabbed and pulled into the Shuttle's payload bay before being fixed. The Hubble Space Telescope has been captured, repaired, and upgraded on-orbit three times since its deployment in April 1990.

The space suit, also called an EVA suit for *extra-vehicular activity*, is a life support system. Usually made of 9 or more layers of material, it's the only thing protecting an astronaut from the cold, heat, micrometeorites, and deadly vacuum of space. The suit carries all the water, food, oxygen, communications gear and electrical power that the astronaut will need to do their job outside a spaceship for up to 8 hours.

Underneath the white space suit, an astronaut wears a layer of long underwear with tubes of water running through it. This garment keeps the astronaut cool while working in space. An astronaut can overheat in sunlight, or she can freeze in the shade. From shadow to sunlight the temperature changes from minus 200 degrees F to plus 250 degrees F. The suit protects her from both extremes. NASA's latest space suits can be used about 24 times before they need to be overhauled.

Space suit sizes for smaller women are a challenge. At least seven of NASA's active women astronauts, including Anna Fisher, are too small to comfortably fit into the current space suit design. Fisher says, "There are a lot of women who fit into the next larger suit, but feel they would do better in the small suit."

Because of budget cuts, NASA had to stop work on a space suit for smaller astronauts. Many of the smaller astronauts, both women and men, hope that NASA will get the funds to restart the project.

Not all astronauts have to fit into an EVA suit, only those who plan to spacewalk, or who might do an unplanned spacewalk during their mission. However, residents of the ISS must be qualified to do a spacewalk.

The lack of smaller space suits could have an affect on who is chosen to be an astronaut. It is possible that some highly qualified people could be turned away from the astronaut corps because they are too small to fit in the current space suit design. NASA doesn't want and shouldn't allow this to happen.

Linda Godwin on her STS-108 EVA

Susan Helms takes a spacewalk on mission STS-102

Astronaut Fuel: Eating in Space

Space food has gotten a lot better tasting since *Mercury* Astronaut John Glenn went into orbit in February 1962. He had to eat applesauce from a toothpaste tube. Scientists were worried that people wouldn't be able to swallow in space. They thought that gravity might play a role in swallowing and digesting food. Happily, it just isn't true. Glenn was able to eat and digest his applesauce just fine as he made his three orbits around the Earth.

Other *Mercury* astronauts ate freeze-dried powders, bite-sized chunks, and semi liquids out of aluminum tubes. The astronauts complained about having to eat cuisine that was the consistency of baby food out of the toothpaste tubes, so NASA made more menu changes.

The food improved a lot for the *Gemini* astronauts. Their meals consisted of re-hydrated milk and cubed, semi-liquid, and dehydrated foods covered with gelatin. Some menu items included butterscotch pudding, vegetables and chicken, shrimp cocktail, and of course, applesauce.

Apollo astronauts were even better off. The Moon explorers were the first to get hot water and to use the Spoon Bowl. Food was gooey enough to eat with a spoon from these plastic bowls.

The *Apollo-Skylab* astronauts had a very earth-like eating experience. Footholds allowed them to "sit" at a dining room table and eat breakfast, lunch and dinner. Magnetized trays had cubbies to hold their food packages, which were opened with scissors. They ate with spoons, forks and knives. *Skylab* astronauts were the first to have a refrigerator and freezer with 72 types of food to choose from.

Nowadays, astronauts eat very Earth-like food. All of NASA's space food is made and packaged by a Houston company called SpaceHab. "It has improved dramatically," says Wendy Lawrence. "You get to customize your menu, so it's not that bad."

Rhea Seddon eating on the Shuttle mid deck

Half a year before launch, a Shuttle crew meets with food specialists to taste all of the different types of space food. Then, about five months before launch, the astronauts choose their space menus. All astronauts meet with a nutritionist to make sure that the food they plan to eat in orbit provides them with all the nutrition they need to stay healthy and strong while in space.

Spacewalking astronauts need more calories per day to stay clear-headed and strong during an EVA. For a healthy heart, they need foods high in potassium.

There are several types of food an astronaut can choose from for her on-orbit menu. There is enough room on the Shuttle to carry 3.8 pounds of food per day for each astronaut, not including water, which the Shuttle makes on-orbit with its fuel cells.

Rehydratable Foods: The water is removed from rehydratable foods to make them weigh less and store easily. Foods like powdered fruit punch, coffee and oatmeal are "rehydrated" just before eating, with water created by the Shuttle's fuel cells.

Thermo stabilized Foods: Thermo stabilized foods are heat processed so they can be stored at room temperature in cans and plastic packages. Fruits and fish are stored in cans that open with pull-tabs similar to fruit cups available at any grocery store. Pudding and similar foods are stored in plastic cups.

Intermediate Moisture Foods: Taking some water out, while being careful to leave enough in to maintain the soft texture, preserves foods like dried peaches, pears, apricots, and beef jerky. Intermediate Moisture Foods can be eaten without any preparation.

Natural Form Foods: Ready to eat foods like nuts, granola bars, and cookies are packaged in flexible pouches. Smaller bite-sized items are preferred to avoid crumbs, which can be a big problem in zero-G. Macadamia nuts are a favorite natural form food to eat in space.

Irradiated Foods: Foods like beefsteak and smoked turkey are cooked and packaged in flexible foil pouches and then sterilized by ionizing radiation so they can be kept at room temperature.

Frozen Foods: Quiche, casseroles, and chicken potpie are quick frozen to help keep the food's original texture and fresh taste. If frozen slowly, large ice crystals would form and change the taste.

Fresh Foods: These shelf stable foods aren't processed and have no artificial preservatives. On shorter missions, raw fruits like apples and bananas are taken into space. Tortillas are the bread of choice by astronauts in flight because they don't make crumbs like most bread. One favorite meal is peanut butter and jelly on a tortilla.

Condiments: Astronauts use single-serve pouches of ketchup, mustard, mayonnaise, taco sauce, and hot pepper sauce. Salt and pepper are diluted in water to prevent grains from floating into faces and Shuttle systems.

Refrigerated Foods: To prevent spoiling, foods like cream cheese and sour cream are kept cool.

Wendy Lawrence says that a lot of astronauts like spicy foods while in space. Probably because they relieve the "stuffy head" feeling astronauts get when they first arrive in orbit.

Different astronauts prefer different foods. "My favorite foods are shrimp cocktail, mushroom soup, and tapioca pudding," says Janet Kavandi. "I also really like the Russian stews that I have eaten on *Mir* and on the ISS."

Astronauts eat foods the same way as on Earth. Well, almost the same way. "It is not difficult to eat in space, but you do have to be more careful with your food," says Kavandi. "Liquids escape easily and can float away onto your crewmate's shirt if you're not careful. Small crumbs are also very difficult to contain."

Crumbs can damage the instruments, and become bothersome if they fly into someone's eyes. On Earth, crumbs fall onto the table or floor and are easily cleaned up. In space, crumbs float around until an astronaut vacuums them up or the air vents filter them out. That's why astronauts prefer bite-sized crackers, and tortillas instead of regular bread.

Wendy Lawrence knows what makes a good sandwich in space. "Tortillas are much better than bread. Bread makes crumbs, tortillas don't. They are great for a sandwich. I like my peanut butter and jelly tortillas, they're not bad. We get to fly with real jars of peanut butter and real jars of jelly."

As an example of what an astronaut might eat for a meal, here's Pilot Pamela Melroy's lunch menu for day 10 of STS-92: Thermo stabilized chicken strips in salsa, rehydratable macaroni and cheese, thermo stabilized rice with butter, natural form macadamia nuts, and apple cider.

The shuttle uses fuel cells for electrical power and these fuel cells use hydrogen and oxygen to make power and water. Since the shuttle creates excess water, foods can be brought up in a dehydrated state. This lessens the number of pounds of cargo that must be launched into orbit. This saves NASA a lot of money. Remember that it costs $10,000 per pound to launch something into orbit.

Food on the International Space Station

Astronauts on the International Space Station (ISS) have a larger variety of food to choose from. The galley of the habitation module (HAB) on the ISS stores two weeks worth of food at a time. Food is stored in the freezer, refrigerator, or at room temperature on the Pressurized Logistics Module (PLM). Every two weeks the HAB galley is resupplied.

The ISS uses solar panels to generate its own power. Therefore, there is no excess of water on the ISS like there is on the Shuttle. Water on the ISS comes from stores and is recycled from the cabin air. For the most part, ISS food doesn't need rehydration. Many thermo-stabilized foods are used so that no additional water is needed.

Food on the ISS is heated in a microwave/forced convection oven. A Japanese noodle company is even working on noodles in a cup for space station astronauts.

Sleeping

Astronauts say that sleeping in space is a lot like going camping. You bring your sleeping bag and find a place to hang out near your friends. But in space, you literally do 'hang out.' Most astronauts on the Shuttle and the ISS attach a sleeping bag to a wall, climb in, and fall asleep. Some astronauts choose to sleep in the commanders' or pilots' seat or in one of the four bunk beds available.

Eating in space: The ISS crew eats dinner together, courtesy IMAX

There are two crew compartments available on the ISS. They are just big enough for one person and a sleeping bag to fit into. Each one comes with a great perk - a window to space. Astronauts can watch the world go by as they drift off to sleep.

In her book *To Space and Back*, Sally Ride remembers listening to music and watching the Earth go by as she relaxed before falling asleep on the space shuttle.

Currently, the ISS has a crew of three people and two crew compartments. So, where does the third person sleep? Anywhere in the station they want, as long as they attach themselves to something. It would be dangerous for them to float around and bump into something or someone.

In 2001, Astronaut Susan Helms spent 163 days on the ISS. At bedtime, she had the Destiny Laboratory Module all to herself. She slept on the opposite side of the station from her crewmates who were about 170 feet away. No word on whether her crewmates snored or not.

Wendy Lawrence says that she sleeps well in space as long as she isn't cold. "I would much rather sleep in space. I get a much better nights' sleep in space than I do down on Earth. You can just go inside of that sleeping bag, relax, and it's incredibly comfortable."

Astronauts are scheduled to sleep about 8 hours a night. But, just as on Earth, they may wake up to get a snack or use the bathroom. Unfortunately, some astronauts don't sleep well their first nights in space because of motion sickness. Some nights they may not be able to sleep because of the excitement of being in space. Some astronauts have spent hours looking out the window at the Earth and stars instead of sleeping.

Most astronauts use eyeshades to block out the sun which "rises" every 45-minutes as the Shuttle or station orbits the Earth. Some also use earplugs to drown out the noise of other astronauts, machines, and fans. In case of an emergency, one crewmember wears a communications headset so ground control can call if necessary.

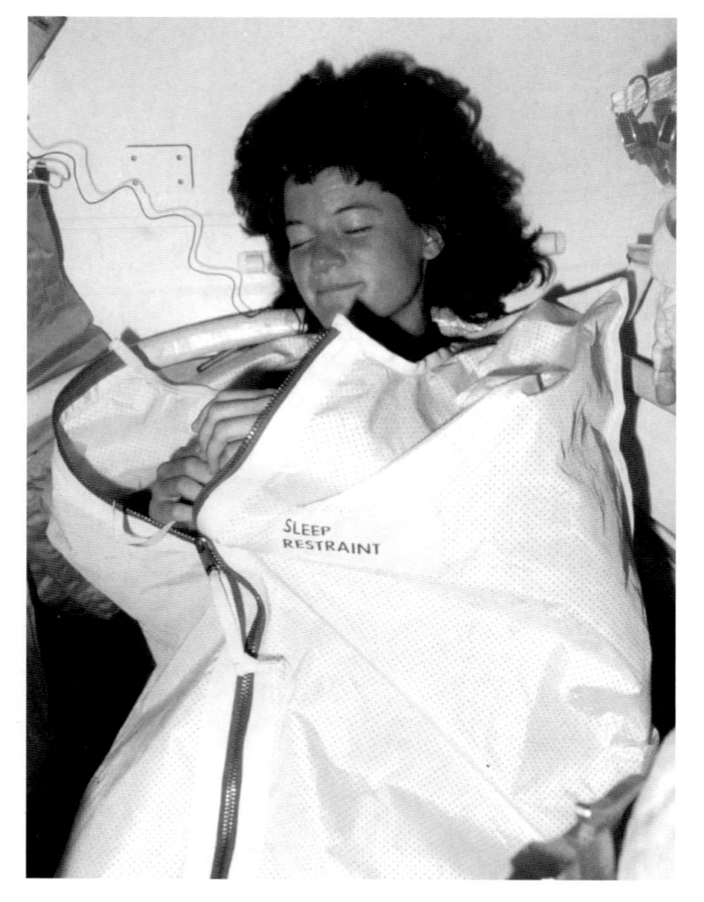

Sally Ride sleeping on her first mission

When relaxing in space, a person's arms tend to float above them. Sally Ride has likened this to a sci-fi movie where zombies move with their arms out in front of them. Some astronauts say that when they fall asleep in space, it's as if their arms don't exist. Astronauts tend to curl into a natural sleeping position in space and stay that way until they wake up. And when they wake up, it's a surprise to see their own arms floating in front of their face. Astronauts dream in space and sometimes even have nightmares. Some astronauts have been known to snore in space, but not on Earth. No one is sure why this happens.

Space shuttle astronauts get a "wake-up call" from Mission Control in Houston every morning of their mission. Usually, the crew's family members choose music that gets the astronauts out of bed in the morning. All kinds of music have been played. The ISS crew uses an alarm clock to wake up in the "morning."

Going to the Bathroom

When nature calls, NASA calls it "digestive elimination." In the movie "2001: A Space Odyssey," Dr. Floyd reads a couple hundred words of instruction on how to use the space toilet. That's pretty close to the truth, but fortunately astronauts train to use their space toilet before their mission. Because of the steps involved, it can take about 10 minutes more to use a toilet in space than on Earth.

Engineers built a toilet for the space shuttle that uses air suction, instead of gravity, to whisk away liquid and solid material. NASA spent $23 million to design and build this space toilet, called the Waste Collection System or WCS. It's probably the most expensive toilet ever built.

Using a toilet in space isn't much different than using one here on Earth. The difference is gravity, or lack thereof. Here on Earth, using a toilet is easy because gravity draws waste down and away from your body. But in space, where there is no gravity, engineers had to find another way to keep astronauts clean and healthy.

Astronauts use an anatomically correct funnel that they hold close to their body to urinate into. Each astronaut has his or her own personal funnel, which is connected to a flexible hose that sucks the urine away. The air that sucks away waste is filtered to remove any bacteria or odor and is then returned to the cabin.

Like a vacuum cleaner, what goes into the WCS is sucked away so that it does not float back. The liquid waste is vented into space and freezes immediately. The rest is stored for disposal when the shuttle returns to Earth. The ISS currently has a Russian-built toilet, which dumps urine overboard and stows solid waste until it can be returned to Earth.

Sit down, strap in, and go to the bathroom. For solid waste, astronauts sit down on the WCS and secure themselves to make sure they don't float off the seat. There are foot loops and other restraints to make the experience more Earth-like.

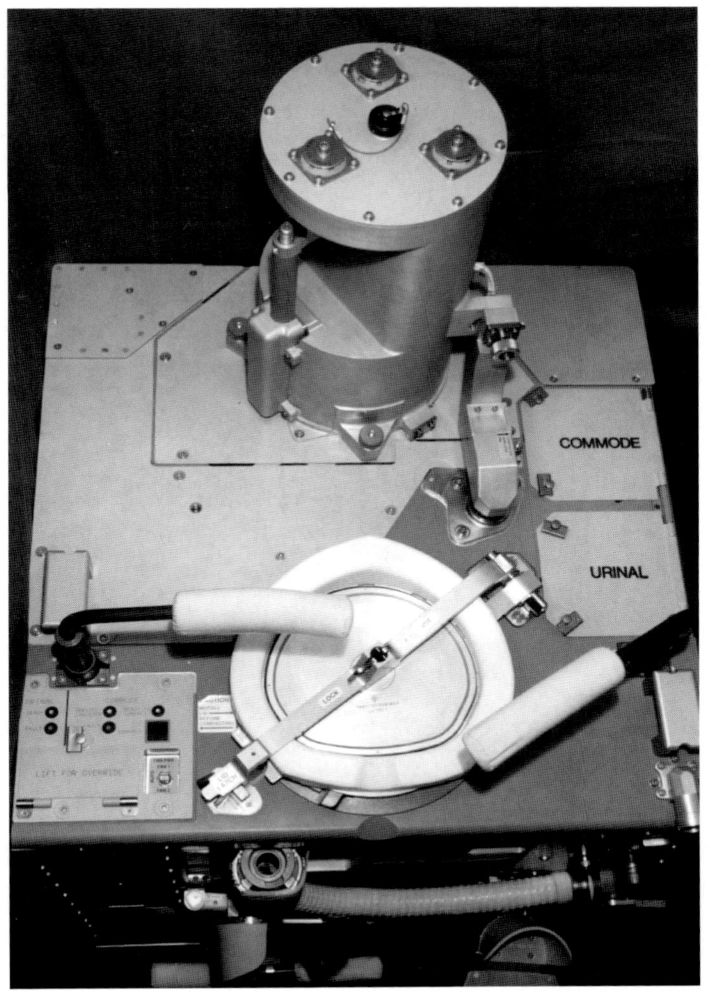

The waste collection system (WCS) in all its grandeur

Anna Fisher says that using the bathroom was the worst part of being in space. However, not all astronauts feel this way. Wendy Lawrence says, "It's really not that hard. It takes a little longer, you have to be a little bit more careful, but it's not that difficult."

"Women have a different type of funnel to catch the urine," says Ellen Baker. "But I don't think it's more of a problem, or more of a difference than here on the ground. Men have an easier time on the ground too, they don't have to disrobe quite as much as women do, and the same is true in space. The bathroom is very easy to use and works pretty well. And generally, when it doesn't work, it's not that it doesn't work for women, it's that it doesn't work for anybody."

Space shuttle toilet problems? How astronauts use the bathroom in space is usually a humorous topic for those of us here on Earth. An October 2000 ABCNews.com headline read "Astronauts Horrified as Toilet Backs Up." However, it's no laughing matter when the space toilet doesn't work. If waste were to float around the cabin, it would be very unhealthy, damaging to instruments, and just plain disgusting.

Astronauts use maximum absorbency garments (called MAGs) during launch, landing and space walks. They are similar to diapers, but much more absorbent. When astronauts go on a spacewalk, they don't come back in until the job is done. So, the unwritten rule is to use the bathroom before putting on your space suit.

Getting Clean

All astronauts have a Personal Hygiene Kit (PHK) with their own toothbrush, toothpaste, comb, etc. Each astronaut gets to decide what goes in the kit and what brands to use.

Without gravity, you can't just pour water from a tap into a sink. The water would go everywhere in clumps and could cause problems. Instead, astronauts get water in slow controlled amounts from a small hose. Astronauts always have to be careful to drink, absorb, or vacuum up any water that they dispense.

Shuttle astronauts use a rinseless body bath and washcloth to take a sponge bath. The concentrated body bath solution is stored in standard drink bags and filled with hot water before use. Most astronauts use rinseless shampoo to clean their hair.

Astronaut Susan Still Kilrain explained to a NASA reporter that it took her a few days to figure out how to use regular shampoo in zero-g. "If I didn't rinse the shampoo out, my hair was very heavy and felt dirty.

So, I ended up washing my hair much like I do at home." She filled a drink bag with water and held a towel over her head as she squeezed the water onto her hair. The towel kept the water from escaping into the rest of the cabin. Then she put shampoo onto her hair and lathered it up. "I found that if I didn't get a lather, I didn't get clean hair," Kilrain remembers. "The tricky part came trying to rinse the shampoo out. I once again used the drink bag to squeeze water onto my hair and with the help of my towel was able to slowly get the shampoo out. Then I'd use my new daily towel to dry my hair a little." Her hair dried quickly in the dry shuttle air. However, "without the help of gravity, there was no controlling my hair once it was washed."

Conclusion

The ceremony of launch is different depending on where you blast-off from. Once in space, it no longer matters. The experience of space is the same regardless of how you got there. Everyone eats, plays in zero-g, and admires the view of Earth. Engineers try to make living in space as easy as possible with clothing, sleeping bags, good food, and hygienic conditions. Hopefully, someday more of us will be able to experience life in space.

2

What Could Have Been: The First Lady Astronaut Trainees

Today, women and men are equals in the astronaut program, but it wasn't always that way. At the beginning of the Space Race, all astronauts were male test pilots. The Soviet Union sent a woman into orbit in 1963, but it would be another 20 years before a woman would be sent into space again. Things might have been very different, however, had the First Lady Astronaut Trainees (FLATS) gotten a chance to show that they too, had the right stuff.

As World War II ended in August 1945, a new battle began between the Soviet Union and the United States of America. It was a war between communism and democracy. The Superpowers were involved in this "Cold War" for most of the second half of the 20th Century. Until the Cold War ended in 1991, nuclear war was a constant threat.

The Cold War gave birth to the Space Age on 4 October 1957, with the launch of Sputnik I by the Soviet Union. The battle for superiority in space had begun.

The Space Race was seen as critically important to winning the Cold War. Vice President Lyndon Johnson wrote in a letter to President John F. Kennedy in 1961, "Other nations ...will tend to align themselves with the country which they believe will be the world leader – the winner in the long run. Dramatic accomplishments in space are being increasingly identified as a major indicator of world leadership."

The Cold War called for cold warriors. Hero's called "Astronauts" who would do single battle against the elements to prove the technological superiority of their nation. In 1959, the first group of candidates for the National Aeronautics and Space Administration's (NASA) *Mercury* astronaut program was tested for physical fitness at the Lovelace Clinic in New Mexico. These men would be the gladiators chosen to represent the United States in the fight to get to space first.

The *Mercury* Astronauts: Alan B. Shepard, Virgil "Gus" Grissom, Gordon Cooper, Walter Schirra, Donald K. "Deke" Slayton, John Glenn, Scott Carpenter

But, what about women? They could be astronauts too. The first nation to put a woman in space would win a minor victory in the Cold War. In the United States, a group of women were briefly considered for astronaut training. It's a part of American history that hasn't gotten much attention. And when it has, the story is often misleading.

Woman in Space Earliest

In September 1959, the Air Force ARDC (Air Research and Defense Command) started the Woman in Space Earliest (WISE) program in an attempt to beat the USSR. Those in the Air Force who started the program felt that the "propaganda victory" of being the first nation to put a woman in space was worth the effort.

Soon after the WISE program began, the Air Force tested 58-year old aviation pioneer Ruth Nichols with a set of aero-medical and psychological tests. She experienced near weightlessness, isolation, and centrifuge testing and passed all the tests. Nichols told a New York Times reporter about the centrifuge tests, "Well, it's the same thing as what I said about being nauseated in an airplane: if you are busy doing something, your mind is completely taken up with that. Therefore you don't think about the sensations of flying."

Also tested late in 1959 was aerobatic pilot Betty Skelton. She went through a set of rigorous physical testing and spaceflight simulations. Unfortunately, her testing was just a publicity stunt for *Look* magazine. The cover of the magazine's February 1960 issue showed Skelton dressed in a silver space suit with the article title, "Should A Girl Be First In Space?"

Dr. Randolph Lovelace II created a program for the Air Force to test *Mercury* astronaut candidates. So, when the Air Force wanted to test women astronaut candidates, they naturally went to him again. Late in 1959, Dr. Lovelace began working with Air Force Brigadier General Donald Flickinger, to decide which qualified women to invite for testing.

The accomplished aviator Jerrie Cobb was the first woman selected for testing, and helped General Flickinger by suggesting other women aviators. However, in December 1959, before any of the women could be formally tested, the Air Force abruptly canceled the WISE program.

To this day, it is still unknown exactly why the Air Force canceled the WISE program. The newly founded NASA didn't yet have it's own aerospace medical department, and relied on the Air Force to provide medical services. Perhaps someone at NASA got wind of the Air Force WISE program and ordered it canceled. Or, perhaps the Air Force didn't want NASA to have women astronauts.

One of the clues is a 7 December 1959 letter from General Flickinger to Jerrie Cobb. "The unfortunate Nichols release did much to turn the tide against Air Force medical sponsorship of the program, and to this day I cannot find out the individual responsible for approving the release," Flickinger wrote.

What was the "unfortunate Nichols release?" That still remains a mystery. From all accounts, the testing of Ruth Nichols was a success. She passed all the tests. The letter implies that other NASA and Air Force leaders didn't know about General Flickinger's WISE program. Apparently, when Nichols's test results were made known, WISE came under harsh scrutiny by Air Force officials who were opposed to women becoming astronauts. This seems to be the real reason why the WISE program was cancelled.

The second clue is a memo from General Flickinger to Dr. Lovelace transferring all of the WISE files to the Lovelace Foundation. In that letter, Flickinger says there was, "too big a chance of adverse publicity to warrant continuation of the project." This memo implies that *someone* thought that there would be public outcry over women astronauts. But, *who* thought there would be an outcry?

The FLATS

Despite the cancellation of the WISE program by the Air Force, Dr. Lovelace decided to press on with the program privately. Lovelace had already lined up Cobb for the astronaut examinations and wanted to test more women. He created the Women in Space program, later called the First Lady Astronaut Trainee or "FLATS" program, to prove that women were just as qualified as men to be astronauts. It was hoped that once NASA saw the results, the space agency would become convinced that women had a place in NASA's astronaut program.

Jerrie Cobb in training

In August 1960, Dr. Lovelace announced Cobb's test results at the Space and Naval Medical Congress in Stockholm, Sweden. Lovelace declared, "We are already in a position to say that certain qualities of the female space pilot are preferable to those of her male colleagues."

Shortly after Dr. Lovelace announced his results, *Life* magazine and the *New York Times* published articles about Cobb. The *Life* article was positive and claimed that Cobb was, "The first prospective space pilot in a hitherto unannounced 12-woman testing program." This article made the program sound official, which it wasn't.

However, a number of other articles were not so open-minded. *The Washington Star*, and *Time Magazine* reduced Cobb to a feminine object instead of the talented pilot that she was. Rather than enlighten the public about her many aviation accomplishments, both publications described Cobb's looks and quoted her weight and measurements. The portrayal of "Bachelor-girl Cobb" is an example of the gender stereotypes of the time that would eventually cause the men in power to kill the program.

In November 1960, Dr. Lovelace asked the wealthy aviator Jacqueline Cochran for funding and support to test more women. Cochran was a leading woman pilot, aircraft racer, and holder of several aviation records since the 1930's. As a member of the Collier Trophy selection committee, Cochran had been largely responsible for bringing Dr. Lovelace's advances in aviation medicine to the public's eye. She had helped him win the Collier trophy in 1940.

Using private funds from Cochran to pay for testing expenses, Dr. Lovelace tested more women for the Women in Space program in December 1960 and early 1961. Over the next 9 months Dr. Lovelace tested about 25 women using the same medical and fitness exams that he used to test the male *Mercury* astronaut candidates.

Thirteen women aviators passed the initial round of tests and proved that they were just as much astronaut material as the *Mercury* men. They were: Myrtle K. Thompson Cagle, Geraldyn "Jerrie" Cobb, Jan Dietrich, Marion Dietrich, Mary Wallace, "Wally" Funk II, Jane Hart, Jean Hixson, Gene Nora Stumbough Jessen, Irene Leverton, Sarah Lee Gorelick Ratley, Bernice B. Steadman, Geraldine "Jerri" Sloan Truhill, and Rhea Hurrle Woltman.

Three phases of testing began with the same examinations that the male *Mercury* astronauts were put through. Doctors wanted to see just how much the human body could stand. In addition to the standard physical exam, many other tests were done including over a dozen types of eye tests. Some of the stranger tests involved swallowing a tube to monitor digestion, injecting dye to measure liver absorption rates, monitoring how fast the hand closed when subjected to electrical stimulation, and a vertigo test that involved squirting freezing water into the ear. Sensory deprivation testing was also conducted in the first phase of exams, even though it was more of a psychological test. The only change to the *Mercury* astronaut testing was the addition of a standard gynecological exam.

The physicals were very thorough and took about a week to complete. "We did not have any secrets when we got out of there," Jerri Sloan Truhill told Historian Margaret Weitekamp, in her book, *The Right Stuff, the Wrong Sex: The Lovelace Woman in Space Program, 1959-1963*.

Much of this physical testing was done because doctors didn't know what zero gravity would do to the human body. They worried about serious problems with the eyes, breathing, balance and digestion. These fears were quickly laid to rest after the first human spaceflight. Later astronauts were grateful not to be subjected to these "torture tests."

The second phase of examinations included psychological tests. These involved isolation in a claustrophobic chamber. In one test, doctors watched for reactions to long periods of boredom followed by short periods of high stress. Only Jerrie Cobb, Wally Funk, and Rhea Hurrle actually took the isolation tests. They all passed.

The third and final phase of testing required using simulators and jet aircraft to simulate the high G's of launch and the zero G's of space. This testing was conducted at the U.S. Naval Aviation Medical Center in Pensacola, Florida. The Navy was reluctant to test civilian women, but in May 1961 they authorized a one-time-only test of Cobb. She successfully passed the Navy's tests.

Unfortunately, the FLATS program was canceled before anyone else had the chance to take any more tests.

A New NASA

It was a year of change for NASA. Early in 1961, James Webb became NASA Administrator. In May, Alan Shepard became the first American in space. Days later, President Kennedy made a stunning announcement, "I believe that this nation should commit itself to achieving the goal, before this decade is out, of landing a man on the Moon and returning him safely to Earth."

These words gave NASA a new mission and focus. No longer was NASA just running tests and trying to do things first. Now the agency had a goal, to get to the Moon. From now on, only tasks that helped the United States get to the Moon before the U.S.S.R. would be funded. Women astronauts became a side issue that would not directly help the U.S. get to the Moon first.

When Dr. Lovelace tried to make arrangements to test other women in the FLATS program at Pensacola,

Alan Shepard in his *Mercury* capsule

the results were not good. Tests originally scheduled for July, slipped into September as the women trained to be in top physical shape for testing. Two even left their jobs to participate in the phase three testing at Pensacola.

On 12 September 1961, just five days before the scheduled testing at Pensacola, each of the women received a telegram from Dr. Lovelace, "Regret to advise you that arrangements at Pensacola cancelled. Probably will not be possible to carry out this part of program."

It seems the Deputy Chief of Naval Operations, Vice Admiral Pirie, had written to NASA in August 1961 to verify that NASA had an official request, or "requirement" to perform the phase three tests on the women astronaut trainees. NASA's Deputy Administrator Hugh Dryden wrote back, "NASA does not at this time have a requirement for such a program." Since there was no official request from NASA, the Navy felt no obligation to perform the examinations, and canceled the tests.

Centrifuge

The abrupt cancellation of the Pensacola tests was a shock to the women involved. Apparently Dr. Lovelace had made only informal arrangements, and had never received an official Navy endorsement of the program. When Dr. Lovelace tried to reschedule, the Navy wouldn't budge, even when he promised to give NASA the test results, "at no cost to the government."

Wally Funk, one of the women astronaut trainees, later managed to undergo phase three testing without the help of the Lovelace Foundation. She arranged to take centrifuge G-force tests and an ejection seat test at the El Toro Marine Base in California. Funk passed both tests.

The women who participated in the testing at the Lovelace Foundation were not promised anything. However, the letters they received, inviting them to apply and informing them that they were selected, implied that it was a more official program than it actually was. The acceptance letter identified the applicant as one of the "potential women astronauts." Nothing incorrect was stated in either letter, there were no promises about what would happen after testing, but the implications were clear. Some of the women left good jobs to come to the Lovelace Clinic for the chance to be an astronaut. Many hopes were dashed when the program collapsed and the lack of government involvement became clear.

While Dr. Lovelace was testing women in 1961, the first astronauts flew into space, proving people could survive spaceflight. In April 1961, Soviet Cosmonaut Yuri Gagarin became the first person in space with a one-hour orbital flight. American Astronauts Alan Shepard and Gus Grissom followed with short sub-orbital flights. These first space flights dispelled fears about what launch and zero-g would do to the human body.

The men who took these first space flights were heroes. The American public wanted to see men who faced danger. They were not ready to accept women as risk takers on dangerous missions representing their country. The United States simply wasn't ready to accept women as astronauts.

A Different Time

In the early 1960's, the United States was a very different place for women than it is now. Women in the military were not allowed to be pilots, or fight in combat units. American society was run by chauvinists who believed it was far too dangerous to send a woman into space. Comments by some NASA officials were down right sexist.

John Glenn said shortly after his historic flight, "I think this gets back to the way our social order is organized really. It is just a fact. The men go off and fight the war and fly the airplanes, and come back and help design, and build and test them. The fact that women are not in this field is a fact of our social order."

Robert Voss, NASA's astronaut training officer said, in 1963, "I think we all look forward to the time when women will be a part of our space flight team. For when this time arrives, it will mean that man will really have found a home in space – for the woman is the personification of the home."

Von Braun with President Kennedy

In 1962, Wernher Von Braun, chief designer of the U.S. rocket fleet, demeaned women as nothing more than playthings for men. In a reply to a question about female astronauts he jokingly said he could, "reserve 110 pounds of payload for recreational equipment."

These statements are perfect examples of what most Americans believed at the time; a woman's place was in the home. Apparently, the only way for a woman to get into space was to be the cook and maid in a man's space house.

Astronaut Shannon Lucid remembers these attitudes were around when she grew up, "When I was in school and I wanted to go to college, the teacher just laughed and asked 'why do you want to waste your parents money, because you're a girl?' When I was studying chemistry, they said, 'there's no point in doing that, go learn how to type so you can get a job.' There was no encouragement for females back in those days."

It's clear that in the 1960's, NASA didn't want women to be astronauts. But the space agency didn't have much choice in the matter, either. For the first astronaut selection groups, NASA only accepted military test pilots. Since women weren't allowed to be military test pilots, then there was no way for them to become American astronauts. Although the space agency was initially considering a wider field of applicants, President Dwight D. Eisenhower told NASA that it must choose test pilots.

Eisenhower had several reasons for making NASA choose test pilots. First, it was a well-known fact that they were gifted pilots. Second, test pilots were used to dangerous missions and knew how to stay calm when things went wrong. Finally, and possibly most importantly, they had already been through extensive security checks. NASA could use them immediately, skipping up to a year of investigations into their backgrounds.

Some of the FLATS were not willing to accept their fate, however. Especially when they knew that their gender was the real reason they would not be traveling into space. As a result of her media exposure, Cobb had taken a consulting position with NASA in May of 1961. At the time, she figured that it would be a positive step towards becoming an astronaut.

Instead, it appears that NASA Administrator James Webb may have appointed Cobb in an attempt to keep her quiet after she had stirred up so much publicity with her Lovelace test results. Having Cobb on staff allowed Webb to claim that NASA was investigating the issue of women astronauts. If his intent was to keep Cobb silent, it failed. Cobb continued to be a vocal advocate for women in space and after a year of being "NASA's most un-consulted consultant" her contract was not renewed.

Congressional Hearings

When Cobb learned of the cancellation of the Pensacola tests, she started a campaign to restart the program. At the capitol, in Washington D.C., Cobb joined with fellow FLATS member Jane Hart, wife of Senator Phillip Hart of Michigan, to push for a reversal of the decision to discontinue testing.

Early in 1962, Cobb and Hart met with Vice President Johnson, who was head of the President's Space Council at the time. They tried to convince Johnson to let the FLATS go though the next phase of testing. Johnson was largely silent during the meeting. After the meeting his secretary wrote a letter for Johnson to send to NASA Administrator Webb. Historian Margaret Weitekamp found this letter at the Lyndon B. Johnson Presidential Library in Austin, Texas. The letter didn't endorse women in space, but asked, "whether NASA had disqualified anyone for being a woman?" Johnson revealed his true feelings when he wrote in large letters at the bottom of the page, "Let's stop this now!" The letter was filed and never sent.

Hart and Cobb didn't give up, however. In July 1962, four months after John Glenn's historic orbital flight, Congress agreed to hold hearings to investigate whether NASA should accept women as astronauts. The House Committee on Science and Astronautics put together a subcommittee to study the controversy.

At a congressional subcommittee hearing in 1962, Hart said, "I am not arguing that women be admitted to space merely so they won't feel discriminated against. I am arguing that they be admitted because they have a real contribution to make."

A *Mercury* Redstone launch

Unfortunately, the arguments Cobb and Hart made did not go over well. NASA's argument was that test pilots were already the most qualified people to be astronauts, and only men were test pilots. Since NASA already had enough male astronaut candidates, there was no need to change the requirements to include women.

Jacqueline Cochran, initially a proponent of the FLATS, drastically changed her attitude, probably because she wasn't given a controlling role in the program. And, it didn't help that she had personal disagreements with Jerrie Cobb. Cochran told the subcommittee that suitable women "should not be searched for by injecting women into the middle of an important and expensive astronaut program." She went on to re-enforce the gender stereotypes of the time, "You are going to have to, of necessity, waste a great deal of money when you take a large group of women in, because you lose them through marriage."

John Glenn's testimony as national hero was probably the most influential to the committee members. His analogy was, "My mother could probably pass the physical exam that they give pre-season for the Redskins, but I doubt if she could play too many games for them."

The committee concluded that, "NASA should continue to maintain the highest possible level of personal qualifications." As a result, test pilots would continue to be a "requirement," and women, who couldn't be test pilots, would continue to be left out of the astronaut program.

Conclusion

Space history might be very different if women had a chance to become NASA astronauts in the early 1960's. Women and minorities would have to wait until 1978 to become American astronauts; after the sexual revolution had taken hold and new laws protecting the rights of women and minorities were put into effect.

Jerrie Cobb was a courageous woman. Historian Margaret Weitekamp points out that, "Cobb was pushing for things that the nation wasn't quite ready for yet."

Most historians see the 1963 publication of Betty Friedan's *The Feminine Mystique* as the beginning of the women's movement. Jerrie Cobb was ahead of her time by attempting to become an astronaut in a male dominated society. She should have been given the chance to become an astronaut, but the gender prejudice of the late 1950's and early 1960's stopped her before she had a chance to make real progress.

The Cold War race to put a woman into space was won not by any of the FLATS, but by the Soviet Union when they launched Valentina Tereshkova into orbit in June 1963. Her flight challenged the American view that women had no business being in space.

None of the women selected and tested for the FLATS program has been to space. But that soon may change. Wally Funk has agreed to buy a $2 million ticket for a seven-day stay in polar low-Earth orbit. Funk will fly on the *Neptune Spaceliner*, a commercial tourism venture from the California-based space rocket developer InterOrbital Systems. The plan is to begin launching paying customers into orbit in 2005 from the Pacific island of Tonga. So, at least one of the FLATS may see her dream of flying into space come true. Godspeed Wally Funk!

3

Women's Historic Place in the Space Age

The Space Race

As the FLATS were trying to become the first women in space, the Soviet Union saw another opportunity to be first. Part of the Cold War mentality was the idea that the first country to do something important was the better, more advanced culture.

Who launched an object into space first? The U.S.S.R did, with *Sputnik I* on 4 October 1957. Who was the first to send a spacecraft to the surface of the Moon? The U.S.S.R., with *Luna-2* in September 1959. Who put the first person into orbit? The U.S.S.R. launched Yuri Gagarin into orbit on 12 April 1961.

Yuri Gagarin, the first person in space, in his capsule.

With all of these "firsts", who would be the first country to put a woman into space? Who would be first to show that they had the more "equal" society, where a woman could do anything that a man could do? Despite the best efforts of the FLATS to convince American leaders, it was the Soviet Union that was first, once again.

The First Woman in Space

On 16 June 1963, Russian parachutist and trained Cosmonaut Colonel Valentina Tereshkova blasted into orbit atop a Soviet rocket and became the first woman to travel into space. She launched into space on the Soviet spaceship *Vostok 6* and spent nearly 3 days in orbit. To this day, Tereshkova remains the only woman to fly solo into space.

After Tereshkova's flight there were several rumors about her performance during the flight. There were claims that she was hysterical and didn't achieve her mission goals. Vassily Mishin, a Soviet Official, claimed, "Tereshkova turned out to be at the edge of psychological stability."

According to Asif Siddiqi, author of the definitive book on the Soviet Union's space program; *Challenge to Apollo: The Soviet Union and the Space Race: 1945-1974,* these rumors were false. Her flight wasn't perfect, but Tereshkova completed most of her mission objectives. She suffered from space sickness for the first day or two. But, so did Gherman Titov, the second cosmonaut in space, and his mission was hailed as a triumph.

Tereshkova didn't manually control the direction of the spaceship on her first orbit as planned because she was denied permission, probably because of her space sickness. On the final orbit of her flight Tereshkova did control the capsule and proved, to those who had doubts, that a woman could pilot a spacecraft.

Her space suit didn't fit well, and Tereshkova wasn't able to do the planned biological experiments

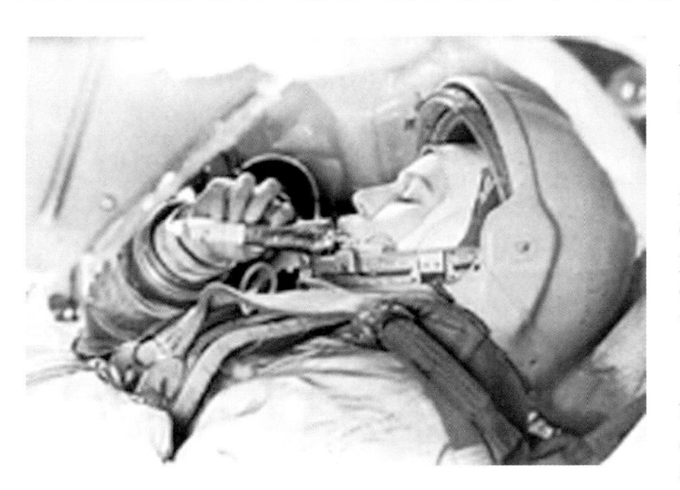

Tereshkova in her capsule

because she couldn't reach them on the other side of the capsule.

After the flight, she gave away her uneaten food to the villagers who gathered around her capsule before the authorities arrived. The physicians were angry because they couldn't determine exactly how much food she had eaten while in space.

These were the only problems on her flight, but they became the basis of rumors that she didn't do her job and fulfill her mission objectives. Gender prejudice and biased mission reports are probably the reason more women cosmonauts didn't get their turn to fly in space.

Soviet Premier Nikita Khrushchev personally selected Colonel Tereshkova to be the first woman in space. But after her flight, other female cosmonauts in her group were never considered for flight assignments on an equal basis with the male cosmonauts. Rather than being integrated into the Soviet space program, women cosmonauts were used for propaganda purposes. The Soviets, it seemed, were almost as prejudiced against sending women into space as the United States.

The Soviet female cosmonaut detachment was officially disbanded in October 1969. It was only in 1980, with the inevitable flight of an American woman astronaut on the space shuttle, that a new group of Soviet women were selected to be cosmonauts.

No woman traveled into space for almost 20 years after Tereshkova's 1963 flight. Throughout the Space Race, astronauts on both sides were white men. The requirements for the job of astronaut emphasized jet pilot skills, and at the time there were few women with much experience flying jets. Certainly not as much as some men with thousands of hours as combat pilots and test pilots.

Saturn V with the American Flag

The Space Race

In America, a few women and minorities applied for the space program and were considered. In 1962, Captain Ed Dwight, a black Air Force test pilot was considered during the second astronaut selection board for the *Gemini* flights. In 1963, two women and four black pilots applied to the third astronaut selection board, but none were selected.

In 1964, the rules for the fourth astronaut selection board were changed. For the first time, NASA opened the astronaut invitation to non-pilots. The space agency received 1,351 applications, which were pared down to 400 finalists. The list of finalists included four women who were considered, but not selected. The six non-pilots selected in June 1965 for the *Apollo* space program were all men; engineers Owen Garriott and Edward G. Gibson, physicist Curtis Michel, physicians Duane Graveline and Joseph Kerwin and geologist Harrison Schmitt.

The race to the Moon heated up when the *Apollo* missions

Earthrise over the Moon

Neil Armstrong steps off the ladder

began. On 21 December 1968, Frank Borman, James Lovell, and William Anders became the first humans to see the lunar surface up close when they orbited the Moon in their *Apollo 8* spacecraft.

The Space Race officially ended when the United States put a man on the Moon on 20 July 1969. "Houston, Tranquility Base here. The Eagle has landed," said Commander Neil Armstrong just after touchdown of Lunar Module *Eagle*.

Armstrong said these famous words when he stepped from the Lunar Module and onto the surface of the Moon, "That's one small step for man, one giant leap for mankind."

Minutes later, U.S. Air Force Colonel and pilot of the Lunar Module Edwin "Buzz" Aldrin joined him on the surface. The astronauts read a plaque to be left on the lunar surface, "Here men from the planet Earth first set foot upon the Moon, July 1969 A.D. We came in peace for all mankind." The *Apollo 11* astronauts and President Richard Nixon had signed the plaque on Earth months earlier.

Geologist Harrison Schmitt was the last man to set foot on the Moon on the *Apollo* 17 mission in December 1972.

NASA's next *Apollo* program venture was *Skylab*, America's first space station. Launched into Earth orbit by a powerful *Saturn V* rocket in May 1973, *Skylab* proved that people could live in space for long periods of time. Astronauts used X-ray telescopes to study the structure of the Sun and did medical experiments to see how the human body adapted to zero gravity. Three crews of three astronauts each visited the space station in 1973 and 1974. *Skylab* burned up in Earth's atmosphere in July 1979.

The Soviets put up their own space stations. *Salyut 4* came first, in 1974 and then came *Salyut 6*, in 1977. Five years later, they replaced *Salyut 6* with the more advanced *Salyut 7* space station.

Footprints on the Moon

The Plaque left on the Moon by *Apollo 11*

The Space Shuttle Era

Before the *Apollo* program was over, NASA began to plan its next great adventure in human space exploration. President Nixon announced NASA's plan for the Space Transportation System in January 1972 and by 19 March, the first Space Shuttle Office opened at Marshall Space Flight Center in Alabama.

The Space Transportation System had been sold to the U.S. Congress as a replacement for all launch vehicles in the United States. It was hoped that the shuttle would be low-cost and the only American spacecraft to put satellites into space. NASA planned for the space plane to launch like a rocket and land like an airplane up to 60 times a year. That's more than once a week! However, these goals proved to be unrealistic because the space shuttle cost more than its creators had thought. Today, it still costs less to launch spacecraft using other expendable launch vehicles like Titan, Atlas, Delta, and Boeing's new Sea Launch.

Only after the *Apollo* and *Skylab* programs ended was the invitation to become an astronaut opened to women and minorities. In July 1976, NASA announced that it was looking for astronauts to fly its newly designed Space Shuttle, "NASA is committed to an affirmative action program with a goal of having qualified minorities and women among the newly selected astronaut candidates. Therefore, minority and women candidates are encouraged to apply."

Each space shuttle mission would need two pilots to control the space plane during launch, orbital maneuvers and landing. Other astronauts would be needed to handle payloads, conduct spacewalks, and do experiments.

The space agency needed a new type of astronaut who could learn how to carry out different mission specific jobs involving the payload, scientific experiments, and engineering tasks. The Mission Specialist was born.

In January 1978, the eighth astronaut selection group chose 35 astronaut candidates out of the 8,079 who applied for the new space shuttle program. Most were non-pilots hired as Mission Specialists. For the first time, the list included six American women astronaut candidates. Sally Ride, Judy Resnik, Kathy Sullivan, Anna Fisher, Rhea Seddon and Shannon Lucid would be the first six American women to fly into space.

NASA's first six women astronauts pose with a spacesuit

These six women were selected because they were the best qualified for the job. The official NASA press release reads, "We have selected an outstanding group of women and men who represent the most competent, talented and experienced people available to us today."

The selection of these six women astronaut candidates was the culmination of changes in American society that took place in the 1960's and 1970's. The Civil Rights Act of 1972 required that all Americans be given equal opportunity regardless of race or sex. Professions that had traditionally been male dominated were starting to attract more women; police, rescue workers, doctors, lawyers, and pilots. The military, where many astronauts were recruited from, began to allow women to fly jets. All of these factors played a role in the inclusion of women and minorities in the astronaut selection class of 1978.

American Women in Space

After one year of astronaut candidate testing and training (ASCAN), the candidates were officially declared astronauts. But before any of NASA's six women astronauts would fly, the Soviet Union, not to be outdone by the United States, put a second woman into orbit. Cosmonaut Svetlana Savitskaya launched into space on *Soyuz T-7* in August of 1982, a full 10 months before the first American woman would fly into space.

NASA chose Astrophysicist Sally Ride to be America's first woman in space for its seventh space shuttle mission, *STS-7 Challenger*. While many saw this as a giant leap forward for womankind, NASA made it known that it saw Ride as a qualified mission specialist who just happened to be a woman.

On the other hand, the media referred to Ride as "Sally" in headlines and in broadcast reports. At a press conference one month before her flight, Ride was asked if her voyage into space would affect her reproductive organs and if she intended to have children after the flight. She was also asked if she cried in the shuttle simulator when something went wrong and if she thought that women should be astronauts. As the first American woman astronaut was preparing for her flight into space, a chauvinistic news media was touting the beliefs of the past.

The Space Race was over, but Cold War rivalries still continued. President Ronald Reagan had recently called the Soviet Union the "Evil Empire" and the U.S.S.R. had recently invaded Afghanistan. Both sides were designing and building elaborate nuclear missiles and aiming them at each other. Reagan talked about building "Star Wars," an anti-missile platform in space, to defend the United States against an attack by the Soviet Union.

Even as the politics of the Cold War played on, progress in space exploration continued. On 18 June 1983 the space shuttle *Challenger* launched on mission STS-7 commanded by Captain Robert Crippen. Onboard were five astronauts, including the first American woman in space, Astrophysicist Sally Ride.

Sally Ride talks with the CAPCOM

Valentina Tereshkova congratulated Ride at a Moscow press conference on 22 June. "I am pleased to know that the third representative of the women of the world, now from the United States, is in orbit today," Tereshkova said. "When we were beginning to explore outer space, our goal was to put this unknown world at the service of the people, at the service of human progress. I believe that we will be unanimous that outer space should remain peaceful for all times, free from any kind of weaponry, and serve the cause of peace and understanding among nations."

The crew of STS-7 deployed satellites for Canada and Indonesia. They also operated the Canadian built robotic arm that performed the first deployment and retrieval exercise with the shuttle Pallet Satellite (SPA-01), and conducted several experiments including the U.S./German cooperative Materials Science Payload (OSTA-2). The mission was a success and landed at Edwards Air Force base in California on 24 June 1983.

After her historic flight Ride told reporters, "I'd love to go into space again, I just wouldn't want to go through training!" But apparently the trip was worth the tough training period, because she did go through it all again and made her second flight into space on STS-41G in October of 1984.

Meanwhile back in the U.S.S.R., Cosmonaut Svetlana Savitskaya was training for her second flight into space, and for her first spacewalk. When asked how she felt about becoming the first woman to walk in space, Savitskaya said, "A hundred years from now, no one will remember it. And if they do, it will sound strange that it was once questioned whether a woman should go into space."

In July 1984, Savitskaya launched on *Soyuz* flight T-12 to the Soviet space station *Salyut 7*. She and crewmember Commander Vladimir Dzhanibekov spent three-and-a-half hours outside the station on a spacewalk. They took turns testing an electron beam gun by cutting through titanium, welding metals together, soldering lead to tin, spraying coatings on to disks, and collecting samples from the outside of the space station. The mission was a milestone in space exploration history not only because of the first spacewalk by a woman, but for demonstrating on-orbit construction techniques.

One month later, on mission STS-41D, American astronaut and electrical engineer Judy Resnik became the fourth woman to fly in space. This was the maiden voyage of the space shuttle *Discovery*. Commanded by Hank Hartsfield, the crew of this seven-day mission deployed three satellites, and conducted a number of experiments including filming the Earth and space on an IMAX camera for the movie *The Dream Is Alive*. The crew overcame icing problems on the arm, or Remote Manipulator System, and earned the nickname "icebusters."

Mission STS-41G *Challenger* was Oceanographer Kathryn Sullivan's first flight into space. The crew of seven, commanded by Robert Crippen, launched on 5 October 1984. While on the eight-day mission, Sullivan and Commander David Leestma successfully conducted a three-and-a-half hour extra-vehicular activity (EVA) to demonstrate satellite refueling in orbit. It was the first spacewalk of an American woman. As she floated into the cargo bay and looked at the Earth floating 140 miles below her, Sullivan excitedly said, "I love it ... that is really great."

Crewmember Sally Ride, now on her second mission, joked with Sullivan and Leestma from inside the orbiter. Leestma asked, "Is it time for lunch yet?" Sullivan replied, "I bet they ate our lunch. To which Ride said, "Hey, you'd have loved it."

In November of 1984, medical doctor Anna Fisher, also from the class of 1978, made her first flight into space on STS-51A *Discovery*. She was the first mom in space. Today, Anna Fisher is the only astronaut from the class of 1978 who is still an active astronaut. After a seven-year leave of absence from 1989 to 1996 to raise her two daughters, Fisher returned to the program. She is now working on various technical assignments and is waiting to be assigned to a Space Shuttle flight or Space Station expedition crew.

Mir docks with *Atlantis* on STS-71. Dunbar and Baker are onboard

Astronaut and surgeon Rhea Seddon was the next woman to launch into space in April 1985 on STS-51D *Discovery*. Seddon wanted to be an astronaut since May 1961, when at age 14, she watched on television, as Alan Shepard became the first American in space. Seddon figured out how to do CPR in space. In 1981, she and Robert Lee "Hoot" Gibson were the first astronauts to marry.

Seddon and her crewmates launched a telecommunications satellite for Canada, and Syncom IV, a U.S. Navy satellite. A malfunction in the Syncom satellite resulted in the first unplanned EVA and proved the value of having a crew of astronauts deploy a satellite rather than use an expendable launch vehicle. Seddon also conducted a number of medical experiments while onboard.

Biochemist Shannon Lucid was the last woman from the class of 1978 to launch into space. Her first mission, STS-51G *Discovery* launched in June 1985. While she was the last woman from her class to fly, Lucid eventually had the most flights with 4 more missions over the next decade.

Until June 2002, Lucid held the record for the most time spent in orbit by an American for her 188-day mission on the Soviet Space station *Mir* in 1996. "It was very interesting," says Lucid. "To a certain degree it was sort of like living in Russia, because I was living with two Russians and *Mir* was built by the Russians, so it had a lot in the way Russians build things."

"It was really neat to be able to live and work in space for the extended period of time because being in space for a longer period of time is different than from just a short flight," says Lucid. "It gave me the opportunity to feel like I was really a citizen of space and not a citizen of Earth."

President Bill Clinton gave Lucid the Congressional Space Medal of Honor in 1996. "She has set a remarkable example for a new generation of young Americans, especially young women." Lucid is the first woman to receive this honor.

All NASA astronaut candidate groups selected since 1978 have included women. Bonnie Dunbar and Mary Cleave were selected in astronaut group nine in January 1980. Dunbar and Cleave made their first flights late in 1985 on missions STS-61A and STS-61B, respectively.

And then came the tragic flight that changed NASA's shuttle program forever.

The Challenger Tragedy

On 28 January 1986, the Space Shuttle *Challenger* exploded 72 seconds after launch, killing all seven astronauts onboard. It was the greatest tragedy the U.S. space program had ever faced. The accident ended the lives of the seven *Challenger* crew members: Commander Frank Scobee, Pilot Mike Smith, Mission Specialists Ellison Onizuka, Ron McNair, and Judy Resnik, payload specialist Greg Jarvis, and the first teacher in space, Christa McAuliffe.

The crew of STS-51-L *Challenger*

The accident was a major blow to NASA and to space exploration supporters all over the planet. Millions of children watched the launch at school that day because Christa McAuliffe was to be the first teacher in space. Sadly, the *Challenger* only reached a height of 46,000 feet, before exploding and killing everyone on board. Astronaut Judy Resnik and Teacher in Space Christa McAuliffe became the first women to die onboard a spacecraft.

In the early 1980's, NASA decided it could raise public awareness and interest, particularly among children, if it sent a teacher into space. After an extensive search of applicants, NASA chose Christa McAuliffe to be the first teacher in space and she began training with her STS-51-L crewmates. Her death in the *Challenger* disaster meant an end to NASA's efforts to send non-astronauts into space.

Things changed for the space shuttle program after the loss of the *Challenger*. Space shuttle launches were put on hold for almost three years while the entire Space Transportation System (STS) was scrutinized from top to bottom.

NASA began an intensive look at the Shuttle with the intent of correcting safety problems. Sally Ride stopped training for her third shuttle mission and became part of the President's commission to investigate the disaster and recommend changes to the Shuttle program.

NASA has made thousands of safety improvements to the space shuttle fleet and to the launch process. Astronauts know the dangers of space travel better than anyone. And any astronaut will tell you that the risk is worth it.

Ride reminds us what the astronauts and everyone that works with NASA understands, "There are risks associated with flight. That was true in the early days of the space program and it's still true today. Rocket technology is risky, and those risks are something that every astronaut has to internally come to grips with and be willing to accept," Ride explains. "One of the roles of the engineers at NASA is to understand the space shuttle and the space station so well as to minimize those risks and to make the risks acceptable for the astronauts and the other people involved. So there are definitely risks associated with it, but they are risks that are pretty well understood and are quite well managed."

While the shuttle program was on hold after the *Challenger* disaster, the Soviets re-established their prestigious place in space with the launch of the *Mir* space station.

The flawless 28 September 1988 launch of the Space Shuttle *Discovery* marked the return to flight of the American space program and the beginning of a new era of space exploration. NASA's mission for human space exploration had changed.

The space agency needed to keep its astronauts as safe as possible. From now on, the Shuttle would carry only those payloads that needed human attention. When the payload could more easily be carried by an unmanned launch vehicle, it was. The Shuttle launched only special payloads like the *Magellan* probe to Venus, the *Hubble Space Telescope*, and the *Galileo* probe to Jupiter.

The Hubble Space Telescope

Repair missions to satellites like the *Hubble* would also require Shuttle missions. Spacelab experiments, tether experiments, microgravity experiments, and radar topography mapping required a person to operate them.

Since the *Challenger* accident, NASA has been hesitant to let observers, tourists, and other non-astronauts, who have no direct mission-related assignments, fly on the shuttle. The death of McAuliffe and her crewmates forced NASA to acknowledge the danger of space flight and make safety changes.

The Russian Space Agency appears to have no problems sending a non-astronaut into space. The Russians believe that if a person knows the risks of space flight and is willing to take them, then why should they stop them.

Space tourist Dennis Tito paid an alleged $20 million to fly to the International Space Station on a Russian launch vehicle in April 2001. At the time, there was some speculation that NASA would relax its "no tourists rule." But, NASA insists that anyone flying on the Shuttle must go through astronaut training.

Dennis Tito was not the first space tourist, merely the first one to pay his own way. In the late 1980's, a British consortium decided to pay the Soviets to take people up on commercial flights. Helen Sharman was one of two British citizens selected to become a space tourist.

Sharman was a chocolate manufacturing engineer who worked for the Mars confectioners company. She answered a radio advertisement to become the first British space tourist. She was selected and trained in Russia for her flight. When the consortium collapsed, the Soviets agreed to finish the program at their own expense. In May 1991, she launched with two Soviet cosmonauts to the *Mir* space station. Once onboard, Sharman performed a number of experiments including how weightlessness affects people, plants, and crystal growth.

In April 2002, the Soviets flew South African millionaire Mark Shuttleworth to the International Space Station. After completing a week of training at Johnson Space Center, Shuttleworth received NASA's blessing for his flight. Perhaps attitudes at NASA are changing once again.

Era of Internationalism

An enormous political change occurred in December 1991. The communist government of the U.S.S.R. fell to the will of the people. By January 1992, Russia and the other states of the former Soviet Union became independent nations. Soon after, the United States signed an agreement with Russia to conduct joint space missions.

In February 1994, Cosmonaut Sergei Krikalov joined the crew of STS-60 and became the first Russian to fly on the U.S. space shuttle. The United States began exchanging astronauts for long duration missions on Mir. Between 1995 and 1998, seven American astronauts, Norman Thagard, Shannon Lucid, John Blaha, Jerry Linenger, Michael Foale, David Wolf, and Andy Thomas, lived on the Russian space station *Mir* learning how to live in space.

Air Force pilot Eileen Collins next to an F-4 Phantom

Women astronauts made it to the next level on 3 February 1995 when NASA Astronaut and U.S. Air Force Colonel Eileen Collins became the first woman to pilot the space shuttle. Mission *STS-63 Discovery* was the first flight of the new joint American-Russian space program. As the pilot, Collins was responsible for maneuvering the orbiter into the right place at the right time. In addition to a rendezvous with Mir, STS-63 included, a spacewalk, SpaceHab experiments, and the deployment and retrieval of an astronomy satellite.

As if completing a circle, Collins invited all of the FLATS to attend her launches. Several did. So far, Jerrie Cobb has attended all of Collins's launches. Jerri Sloan Truhill, another of the FLATS, told National Public Radio that the 1995 Collins launch was, "... just about the most moving thing, I believe, I've ever experienced. She took something from each of us into space, except me," said Truhill. "I told her that she was carrying my dreams, that was all that was necessary."

Americans aren't the only astronauts who get to fly on the space shuttle. Astronauts from several countries have flown as mission specialists or payload specialists. Several women are in this group.

One of the first six astronauts chosen by the Canadian Space Agency in 1983, Roberta Bondar was the first Canadian woman astronaut to go into space. She flew onboard STS-42 *Discovery* in January 1992 as a payload specialist. Bondar conducted life science and material science experiments on Spacelab.

The first Japanese woman astronaut, medical doctor Chiaki Naito-Mukai launched on STS-65 *Columbia* in 1994 as the payload specialist for the International Microgravity Laboratory.

Mukai's second flight was STS-95 *Discovery* in 1998. She conducted medical tests on astronaut John Glenn, whose return to space after 36 years was mostly to observe the effects of spaceflight on a senior citizen. In this case, Glenn wasn't only the payload specialist. He *was* the payload. Mukai and other crewmembers tested how the absence of gravity would affect Glenn's metabolism, blood flow, sleep, balance and perception, immune system response, and the density of his bones and muscles.

Russia's third woman astronaut was Cosmonaut Elena Kondakova. On her first flight into space, Kondakova served as Mir's Flight Engineer from 4 October 1994 until 22 March 1995. She was the only woman to live on the *Mir* space station other than Shannon Lucid. She later served as Mission Specialist for a Mir-Shuttle rendezvous onboard *STS-84 Atlantis* in May of 1997.

Claudie André-Deshays Haigneré is a European Space Agency (ESA) Astronaut. She has never flown on the space shuttle, but in a joint mission with the Soviets, she became the first French woman astronaut when she flew to *Mir* on the Russian launch vehicle *Soyuz TM-24* in August 1996. She and her crewmates brought up a lifeboat to the International space station in October 2001 on mission *Soyuz TM-33*.

December of 1998 began the on-orbit assembly of the International Space Station (ISS) with the attachment of the Russian-built Zarya Command Module to the American-built Unity module. The ISS is truly a symbol of international cooperation because several nations contributed to its construction. The ISS finally superceded the *Mir* as the new long duration space station when the *Mir* was de-orbited on 23 March 2001. It was a sad day for those that worked on the great space station that taught Russians and Americans so much about living in space.

ISS and an EVA: Astronauts take a walk in space, courtesy IMAX

A Woman in Command

For 36 years, Russian Cosmonaut Valentina Tereshkova remained the only female commander of a space mission for her solo flight of the *Vostok 6* spacecraft in 1963.

NASA Announced in March 1998 that, for the first time, it had chosen a woman to command a space shuttle mission. On 23 July 1999, Eileen Collins again made history when she became the first woman to command a space shuttle mission with the launch of STS-93 *Columbia*. (Four years earlier, on STS-63 *Discovery*, she had become the first woman to pilot the space shuttle.) At the launch were members of the original FLATS selection team including Jerrie Cobb.

The launch was scrubbed twice. On the third try, a communications problem pushed the launch back 7 minutes. Then, about 5 seconds after liftoff, there was a problem with two of the three main engine backup systems. The main engines continued to work normally and the shuttle reached orbit safely. However, because of a hydrogen leak in the third main engine nozzle, the *Columbia* reached an orbit that was 7 miles short of where it was supposed to be. Collins and her crew maneuvered the *Columbia* into the right orbit and the rest of the mission went almost perfectly. The crew deployed the Chandra X-ray observatory.

The 21st Century

Aeronautical Engineer Susan Helms became an astronaut in 1991 and flew on her first mission, STS-54 *Endeavour*, in January 1993. She was the first female resident of the ISS. Helms has been on five shuttle flights. Following Shannon Lucid's lead, Helms spent five-and-a-half months living and working on the ISS in 2001 as Flight Engineer on the Expedition 2 crew. While on ISS, Helms installed an airlock and took a spacewalk as part of the ISS construction.

Susan Helms works the camera equipment

Starting in June 2002, Peggy Whitson will spend four months as ISS flight engineer. This will be her first flight into space.

Barbara Morgan was the backup Teacher in Space for Christa McAuliffe. In 1998, she was chosen as an astronaut candidate. Morgan will be the first Educator Mission Specialist when she flies to the ISS in 2004. NASA plans to send more educators into space with a new program called the Educator Mission Specialist program.

Today's women astronauts are treated no differently than male astronauts and have the same opportunities, requirements, and duties as their male counterparts. Astronaut Janet Kavandi says, "The best thing about being a female astronaut is that no one treats you like a female astronaut. You are just another member of the team. I like that."

Now that you have some background on women in space exploration, it's time to meet the extraordinary women astronauts.

4

Biographies of the Women of Space

Chronological order by flight date

Valentina Tereshkova (U.S.S.R.)
Svetlana Savitskaya (U.S.S.R.)
Sally Ride
Judith Resnik
Kathryn Sullivan
Anna Fisher
Margaret Rhea Seddon
Shannon Lucid
Bonnie Dunbar
Mary L. Cleave
Sharon Christa McAuliffe
Ellen S. Baker
Kathryn C. Thornton
Marsha S. Ivins
Linda M. Godwin
Helen Patricia Sharman (UK/*Soyuz*)
Millie Hughes-Fulford
Tamara E. Jernigan
Roberta L. Bondar (Canada)
N. Jan Davis
Mae Jemison
Susan J. Helms
Ellen Ochoa
Nancy J. (Sherlock) Currie
Janice E. Voss
Chiaki Naito-Mukai (Japan)
Elena V. Kondakova (Russia)
Eileen Marie Collins
Wendy B. Lawrence
Mary Ellen Weber
Cathryn G. Coleman
Claudie (Andre-DeShays) Haigneré (France)
Susan L. (Still) Kilrain
Kalpana Chawla
Kathryn Hire
Janet Kavandi
Julie Payette (Canada)
Pamela Ann Melroy
Peggy Whitson
Laurel Clark
Sandra H. Magnus
Heidemarie Stefanyshyn-Piper

Future Astronauts (not yet scheduled for flight as of 30 April 2002)

Yvonne Cagle
Tracy E. Caldwell
Joan E. Higginbotham
Nadezhda Kuzhelnaya (Russia)
K. Megan McArthur
Barbara R. Morgan
Lisa M. Nowak
Karen L. Nyberg
Nicole Passonno Stott
Naoko Sumino, (Japan)
Sunita L. Williams
Stephanie D. Wilson

In Memoriam
Patricia Hilliard-Robertson

Astronaut Biography of Valentina Tereshkova

Personal Information:

The world's first woman in space was born on 6 March 1937 in Maslennikovo, a small town in the Yaroslavl Region of Russia.

Valentina's father was a tractor driver who was killed in action during World War II. Her mother worked in a textile plant. Valentina has a younger brother, Vladimir, and an older sister.

When she was young, Valentina was interested in parachute jumping and this experience led her to space flight. Her first jump was at age 22 at the local Aviation Club in Yaroslavl on 21 May 1959. She then set up and led a parachute club at the Textile Mill for the workers. Two years later she became secretary of the local Komsomol (Young Communist League).

Valentina became the first woman to fly into space on 16 June 1963.

After her flight, Valentina married cosmonaut Andrian Nikolayev, the only bachelor to have flown at the time of her flight. It is rumored that Soviet Premier Khrushchev pressured them into the marriage. The ceremony took place on November 3, 1963, at the Moscow Wedding Palace, and the reception followed at a governmental mansion set apart for state receptions. Khrushchev himself presided at the party, along with top government and space program leaders.

On June 8, 1964, Valentina gave birth to a daughter, Yelena Andrianovna, who grew up to become a doctor. Valentina and Andrian Nikolayev were divorced shortly after Yelena's birth.

Valentina re-married Yuri Shaposhnikov a surgeon and the former director of the Central Institute of Orthopedics and Traumatology. "He was an unbelievably gifted, strong, and decent person," She told the Moscow Times. "With him I did the impossible: I achieved happiness."

Valentina has a grandson Alyosha. "He makes me so happy and gives me so much strength and energy. I am always happy to be with him. And of course my daughter, Yelena, is the closest and most dear person to me."

Education:

Valentina Tereshkova began school in 1945 in Yaroslavl. In 1953, at the age of 16, she left school and began working, but continued her education by correspondence course. By 1961 she had earned a certification as a cotton-spinning technology expert. After her space flight she enrolled in the Zhukovskiy Military Air Academy, graduating with distinction in 1969.

Experience:

Valentina Tereshkova was one of five women cosmonaut candidates picked by the Soviet Union in 1962. Her training included weightless flights in aircraft, flight lessons in jet fighters, parachute jumps, isolation tests, centrifuge tests, and learning about the *Vostok* system.

According to "Who's Who in Space," Tereshkova recounted that some of her male cosmonauts avoided her during training, "because I have invaded their little playground and because I am a woman."

Tereshkova is now the head of the Russian Center for International Scientific and Cultural Cooperation. The center acts as an intermediary assisting research organizations and businesses in establishing contacts between Russia and the rest of the world. "I enjoy introducing foreign governments to our achievements in the areas of culture, science, and technology," she told the Moscow Times. "And even now, despite everything, there are many such achievements. Our people are unusually gifted and talented."

Space Flight Experience:

Valentina Tereshkova became the first woman in space aboard *Vostok 6*. Soviet controllers on the ground were unhappy with Tereshkova's performance in orbit and she was not permitted to take manual control of the spacecraft on the first day, as planned. This was because she suffered from space sickness, which is not uncommon. Several male cosmonauts had the same problem on their space flights. She did take control of the spacecraft attitude later in the flight.

Tereshkova had tasks assigned to her for the first day of flight only. When the flight was extended for a second, and then a third day, there was nothing for her to do but look out the window. Even though Ground control apparently did nothing to support her during that time, Tereshkova never objected and did everything she could to complete the flight.

After Tereshkova landed, local farmers arrived at her capsule well before officials could, so she gave them the food she had left over. This apparently upset a Soviet experiment to monitor how much food she had consumed during her mission.

It appears that Soviet officials exaggerated Tereshkova's flight problems. They claimed she experienced overwhelming emotions, tiredness, and a sharply reduced ability to work and complete all of her assigned tasks. This characterization was unfair and based entirely on her space sickness, few tasks, limited ground support, and her accidental disruption of a food consumption experiment.

The Moscow Times asked Tereshkova about the rumors that she had been taken out of her capsule ill and half dead. She replied resentfully, "That is absolute nonsense, one of the stupid and ridiculous rumors that has been spread about me. After all, my condition, every move I made during the flight was recorded... I carried out all my tasks. And if I was 'half-dead' is it really possible that I could have done that?"

Tereshkova was in space for 3 days. That's more than all the American astronauts at the time, put together. She remains the only woman ever to have flown solo in space.

Special Honors:

After her 1963 flight, Tereshkova received the Order of Lenin and was made a Hero of the Soviet Union. She received an honorary commission in the Soviet Air Force.

Although she never flew in space again, Tereshkova became a spokeswoman for the Soviet Union's space efforts. She has received the United Nations Gold Medal of Peace, the Simba International Women's Movement Award, an additional Order of Lenin, and the Joliot-Curie Gold Medal.

London's International Women of the Year Association named Tereshkova the Woman of the Year in 1984. The organization gave her its highest honor when they named her Woman of the Century in 2001.

Astronaut Biography of Svetlana Yevgenyevna Savitskaya

Personal Information:
Born 8 August 1948 in Moscow, Russia.

As a child Svetlana enjoyed reading, figure skating and running. At her parent's insistence, she took up music, English lessons and swimming.

Svetlana's father Yeveniy Savitsky, was the Deputy Commander of the Soviet Air Defenses, a World War II air ace, and twice Hero of the Soviet Union. She probably owes part of her success to his influence as well as her own natural flying ability and perseverance.

At age 15, Svetlana forged her birth certificate to say she was older so that she could make a solo airplane flight. Her father pretended not to know about it, but went to the airfield to see her flight anyway. He was proud of her and greeted her with a chocolate bar when she landed.

By the age of 16, she began parachute training. When she was 17, her father supported her attempt to set a stratospheric sky dive record. She jumped from 46,758 feet and fell for 45,000 feet before opening her parachute at 1600 feet.

Svetlana is married to Viktor Khatkovsky, an engineer and pilot at the Ilyushin Aircraft Design Bureau.

She enjoys the music of Beethoven and Tchaikovsky, and her favorite poet is Mikhail Lermentov.

Education:
Svetlana Savitskaya is an Engineering graduate of the Moscow Aviation Institute, 1972.

Experience:
Svetlana had completed 450 parachute jumps by her 17th birthday in 1965. By the age of 18, she began pilot training at the premier Soviet Aviation Engineering School, part of the Moscow Aviation Institute (MAI). By age 20, she soloed in a YaK-18 trainer.

In 1970, before she was out of college, she became a World Champion Member of the Soviet National Aerobatics Team. She competed at the World Aerobatics Competition in Hullavington, England. By 1972, she graduated from MAI and became a flying trainer at DOSAAF (Central Technical Flying School of the U.S.S.R. Voluntary Society for the Promotion of the Army, Air Force, and Navy). After training others to fly, Savitskaya became a test pilot in 1975. She started work at the Yakovlev Design Bureau in 1976 and established many world records in turbo-prop and supersonic aircraft. One of those records was a female speed record of 1667 miles per hour record in a MiG-21 aircraft. She is qualified as a pilot for 20 different types of aircraft.

Astronaut Experience:
Savitskaya was selected as a cosmonaut in 1980 because of her skill and experience. However, there is an element of

Soviet pride here. Not always happy to have female cosmonauts, the Soviets wanted to upstage the upcoming flights of several female astronauts on the American space shuttle. Savitskaya was the key.

In 1982, nearly 10 months before Sally Ride became the first American female astronaut in space, Savitskaya became the second woman in space. On her second mission, in July 1984, Savitskaya became the first woman to walk in space.

Her planned flight as commander of an all-female crew to *Salyut 7* on the occasion of International Woman's Day was cancelled due to the upcoming birth of her baby as well as problems with the space station.

While still on the books as a cosmonaut, Svetlana worked as a Civil Engineer at Energia. In 1987, she was made Deputy to the Chief Designer at Energia.

In 1989, Savitskaya became a Member of Parliament and officially left the cosmonaut corps in 1993.

Space Flight Experience:

Savitskaya has flown on two spaceflight missions for a total time of 19.71 days, including one EVA for 3.58 hours.

Soyuz T-7 launched on 19 August 1982. The *Soyuz* docked with space station *Salyut 7*. On board were Svetlana Savitskaya, Leonid Popov, and Alexander Serebrov. The crew conducted scientific research and technical experiments. The spacecraft returned the crew of *Soyuz T-5* to Earth. The mission duration was 7.91 days.

Soyuz T-12 launched on 17 July 1984 and landed 11 days later. This mission was also to the space station Salyut 7. Savitskaya was the Flight Engineer on this flight commanded by V.A. Dzhanibekov. The crew also included Cosmonaut I.P. Volk who conducted scientific and technical studies and experiments. Savitskaya and Dzhanibekov took a 3.5-hour spacewalk.

Astronaut Biography of Sally K. Ride

Personal Information:

Born 26 May 1951, in Los Angeles, California.

Dr. Sally K. Ride has been called an "American hero", a "hero to women", a "symbol of equal opportunity" and a "role model for women". She believes that women can do anything, and of course she's right!

She grew up watching Mission Control and Moon landings. When Sally was 10, she watched, as Yuri Gagarin and then Alan Shepard became the first people in space. "For as long as I can remember I was interested in space, interested in the planets."

As a girl Sally liked books, especially mysteries and science fiction. "I read all the Nancy Drew books. I read all the Danny Dunn books. I don't know if you've seen those? One was *Danny Dunn and the Anti-Gravity Paint*."

Sally says that playing sports gave her a good appreciation of what it's like being a member of a team. *"I think that one thing that people don't appreciate is how important teamwork is as an astronaut. Astronauts have to be able to work well in groups.* And they have to be able to understand what their role is on a particular mission, and what everybody else's role is on a particular mission, how mission control contributes to the operation of a mission. So the whole concept of teamwork is really critical."

Sally played all kinds of sports, but especially liked tennis. "When I was a girl I wanted to be either a professional tennis player, or a scientist. I didn't really know what kind of scientist. I thought maybe an astronomer, I thought maybe a biologist. This is when I was maybe eight or nine. I didn't really formulate that I wanted to be a physicist until I was just finishing high school and going to college."

When she was finishing up her Ph.D. in Physics at Stanford, Sally saw an advertisement for astronauts. "I was a graduate student a few months away from getting my Ph.D. and NASA had put an ad into the Stanford student paper through the center for Research on Women at Stanford, saying they were accepting applicants for astronauts and they wanted applications from women as well as from men."

Sally was selected and her plans of being a research physicist took a different path. When she became an astronaut, she says her father was the happiest of all. He could finally explain in plain English what his daughter did for a living.

Sally says her choices early in life allowed her to become an astronaut. *"I think the most important choices when I was going through middle school and high school and college were to focus on my interests in science.* Because I was lucky in the sense when NASA made the announcement that they were looking for new astronauts, I had the background and the preparation to apply. That's because I'd pursued those interests in science while I was growing up."

After she left NASA Sally reportedly chewed out the press for making "such a big deal" about a woman being an astronaut. She said, "It's time that people realize that women in this country can do any job that they want to."

NASA administrator James Fletcher said about Dr. Ride, "Her flight as the first American woman in space firmly established an equal role for women in the space exploration program. Today, the assignment of women to shuttle crews is a routine matter based on ability and need and is no longer a cause for notice."

Ride has been able to use her fame to help others. "When young girls who are growing up, and young women who are interested in the space program, and professional women, think of female astronauts, they may think of several of us, but they usually think of me included in that," says Sally. "I think that's led to opportunities to have an impact and to have an influence in education and in science that I wouldn't have had otherwise."

She enjoys tennis (having been an instructor and having achieved national ranking as a junior), running, volleyball, softball & stamp collecting (mostly Olympic Stamps).

Education:

Sally Ride graduated from Westlake High School, Los Angeles, California, in 1968. She received from Stanford University a bachelor of science in Physics and a bachelor of arts in English in 1973, and Master of Science and doctorate degrees in Physics in 1975 and 1978, respectively. "Most of my research, including when I was in graduate school and afterwards, was in laser physics," says Dr. Ride.

Experience:

NASA selected Dr. Ride as an astronaut candidate in January 1978. In August 1979, she completed a 1-year training and evaluation period, making her eligible for assignment as a mission specialist on future Space Shuttle flight crews. She subsequently performed as an on-orbit capsule communicator (CAPCOM) on the STS-2 and STS-3 missions.

Ride has a passion for educating people about science and, since leaving NASA, has written four books about space. Her first book "To Space and Back," is about her experiences as an astronaut on the space shuttle. Her other books are, *"Voyager: An Adventure to the edge of the Solar System", "The Third Planet: Exploring the Earth from Space", "The Mystery of Mars."*

In 1989, Dr. Ride became a physics professor at the University of California, San Diego and Director of the California Space Institute, a research institute of the University of California.

Ride works to make science and technology fun for girls. She has started a company called "Imaginary Lines" which operates the "Sally Ride Science Club". The club for upper elementary and middle school girls, holds festivals, events and even a special Space Camp session aimed at keeping girls engaged in the fun of math and science.

Space Flight Experience:

On June 18, 1983, 32-year-old Astrophysicist Dr. Sally Ride became the first American woman to fly into space. The Space Shuttle *Challenger* launched on a 5-day mission, put a satellite into orbit, and landed at Edwards Air Force base in California.

Dr. Ride was a mission specialist on STS-7, which launched from Kennedy Space Center, Florida, on June 18, 1983. She was accompanied by Captain Robert L. Crippen (spacecraft commander), Captain Frederick H. Hauck (pilot), and fellow mission specialists Colonel John M. Fabian and Dr. Norman E. Thagard. This was the second flight for the Orbiter *Challenger* and the first mission with a 5-person crew. During the mission, the STS-7 crew deployed satellites for Canada (ANIK C-2) and Indonesia (PALAPA B-1); operated the Canadian-built Remote Manipulator System (RMS) to perform the first deployment and retrieval exercise with the Shuttle Pallet Satellite (SPAS-01); conducted the first formation flying of the orbiter with a free-flying satellite (SPAS-01); carried and operated the first U.S./German cooperative materials science payload (OSTA-2); and operated the Continuous Flow Electrophoresis System (CFES) and the Monodisperse Latex Reactor (MLR) experiments, in addition to activating seven Getaway Specials. Mission duration was 147 hours before landing on a lakebed runway at Edwards Air Force Base, California, on June 24, 1983.

Dr. Ride served as a mission specialist on STS 41-G, which launched from Kennedy Space Center, Florida, on October 5, 1984. This was the largest crew to fly to date and included Captain Robert L. Crippen (spacecraft commander), Captain Jon A. McBride (pilot), fellow mission specialists, Dr. Kathryn D. Sullivan and Commander David C. Leestma, as well as two payload specialists, Commander Marc Garneau and Mr. Paul Scully-Power. Their 8-day mission deployed the Earth Radiation Budget Satellite, conducted scientific observations of the earth with the OSTS-3 pallet and Large Format Camera, as well as demonstrating potential satellite refueling with an EVA and associated hydrazine transfer. Mission duration was 197 hours and concluded with a landing at Kennedy Space Center, Florida, on October 13, 1984.

In June 1985, Dr. Ride was assigned to serve as a mission specialist on STS-61-M. She stopped mission training in January 1986 in order to serve as a member of the Presidential Commission on the Space Shuttle *Challenger* Accident. Upon completion of the investigation she was assigned to NASA Headquarters as Special Assistant to the Administrator for long range and strategic planning. Dr. Ride left NASA in 1987.

Astronaut Biography of Judith A. Resnik

Personal Information:

Born 5 April 1949, in Akron, Ohio.

Judy was killed when the Space Shuttle *Challenger* exploded 72 seconds after launch on 28 January 1986.

Judy didn't consider becoming an astronaut until graduate school. "I didn't think about being an astronaut my whole life," she told Maryland Today. "I heard that NASA was looking for astronauts right before I completed my Ph.D. It sounded like a very interesting thing and I'll take my chance."

Judy was a biomedical engineer at the National Institutes of Health while she was getting her doctorate. She didn't know what she would do after she graduated. "I really didn't have any long range goal."

On her first flight in August 1984, while demonstrating microgravity for the camera, she wrote the message "Hi Dad."

"The space shuttle isn't doing a lot of exploration," Judy said. But orbital experiments are just as important. "Going to the moon was doing some exploration, but the shuttle program involves biomedical research which can culminate in better vaccines, metallurgy type of experiments that can result in better alloys, stronger materials, better optics, better electronics, and components that all play back into everything we use on a day-to-day basis.

Judy's advice to students was, *"Go into a field that you enjoy. Study things that are of interest to you and don't try to gear your life to being one particular thing*. There are very few people here [at NASA] who geared their lives to become astronauts and there are a lot of people who have tried to gear their lives to become astronauts and who have not been selected as such."

She went on to advise those who want to be astronauts. *"Having an interest in diversity helps. If you're interested in one field only or one particular facet of one field, you're probably not going to be very happy as an astronaut because we do a lot of different things in a lot of different areas."*

She was the first Jewish astronaut, but wanted to be known for being an astronaut.

Judy was a classical pianist who told *Time* magazine "I never play anything softly". She was a gourmet cook who also enjoyed bicycling, running, and flying during her free time. "I want to do everything there is to be done," she once said.

Education:

Judy Resnik graduated from Firestone High School, Akron, Ohio, in 1966. She received a Bachelor of Science degree in Electrical Engineering from Carnegie-Mellon University in 1970, and a doctorate in Electrical Engineering from the University of Maryland in 1977.

Experience:

Upon graduating from Carnegie-Mellon University in 1970, she took a job with RCA in Moorestown, New Jersey; and in 1971, she transferred to RCA in Springfield, Virginia. Her projects as a design engineer included circuit design and development of custom integrated circuitry for phased-array radar control systems; specification, project management, and performance evaluation of control system equipment; and engineering support for NASA sounding rocket and telemetry systems programs. She wrote a paper about design procedures for special-purpose integrated circuitry.

Dr. Resnik was a biomedical engineer and staff fellow at the Laboratory of Neurophysiology at the National Institutes of Health in Bethesda, Maryland, from 1974 to 1977, where she performed biological research experiments concerning the physiology of visual systems. Before NASA selected her to become an astronaut in 1978, she was a senior systems engineer in product development with Xerox Corporation at El Segundo, California.

Astronaut Experience:

Selected as an astronaut candidate by NASA in January 1978, she completed her 1-year of training and evaluation period in August 1979. Dr. Resnik worked on a number of projects in support of the space shuttle's orbiter development, including experiment software, the Remote Manipulator System (RMS), and training techniques.

Space Flight Experience:

Dr. Resnik first flew as a mission specialist on STS 41-D, which launched from the Kennedy Space Center, Florida, on August 30, 1984. Spacecraft commander Hank Hartsfield, pilot Mike Coats, fellow mission specialists, Steve Hawley and Mike Mullane, and payload specialist Charlie Walker accompanied her. This was the maiden flight of the orbiter *Discovery*.

During this 7-day mission the crew successfully activated the OAST-1 solar cell wing experiment, deployed three satellites, SBS-D, SYNCOM IV-2, and TELSTAR 3-C, operated the CFES-III experiment, the student crystal growth experiment, and photography experiments using the IMAX motion picture camera. The crew earned the name "Icebusters" for successfully removing hazardous ice particles from the orbiter using the Remote Manipulator System. STS 41-D completed 96 orbits of the earth before landing at Edwards Air Force Base, California, on 5 September 1984. With the completion of this flight she logged 144 hours and 57 minutes in space.

Dr. Resnik was a mission specialist on STS 51-L, which launched from the Kennedy Space Center, Florida, at 11:38 EST on 28 January 1986. The crew on board the Orbiter *Challenger* included the spacecraft commander, Mr. F.R. Scobee, the pilot, Commander M.J. Smith (USN), fellow mission specialists, Dr. R.E. McNair, and Lieutenant Colonel E.S. Onizuka (USAF), as well as two civilian payload specialists, Mr. G.B. Jarvis and Mrs. S. C. McAuliffe. The STS 51-L crew was killed when the *Challenger* exploded after launch.

Organizations:

Dr. Resnik was a member of the Institute of Electrical and Electronic Engineers (IEEE), American Association for the Advancement of Science, IEEE Committee on Professional Opportunities for Women, American Association of University Women, American Institute of Aeronautics and Astronautics (AIAA), and a Senior Member of the Society of Women Engineers.

Special Honors:

Dr. Resnik received a graduate Study Program Award from RCA in 1971, and an American Association of University Women fellowship, 1975-1976. She also received a NASA Space Flight Medal in 1984.

Astronaut Biography of Kathryn Sullivan

Personal Information:

Born 3 October 1951 in Paterson, New Jersey, but considers Woodland Hills, California, to be her hometown.

On her first mission, STS-41G *Challenger*, Kathy became the first American woman to do an EVA (spacewalk). She and David Leestma spent 3.5 hours demonstrating satellite refueling in space. As she looked out the cargo bay doors at Earth she said, with typical understatement, "That is really great."

Growing up, Kathy was interested in exploration. *"That's why I went into geology in college,"* she told the Center of Science & Industry in Columbus, Ohio. "It was my way of exploring our universe."

She compared the feeling of being in space to what Olympic athletes go through just before a competition. "We watch them on TV and try to imagine how neat it must be, but once you're there it's all about performance and preparation. There are moments you can sit back and take it all in, but there's stuff to be done, and that definitely takes a focus."

Kathy is a private pilot, rated in powered and glider aircraft. Kathy also enjoys squash, bicycling, backpacking, and reading in her spare time.

When asked which trip she'll remember most from her travels on Earth and into space, Kathy says, "That's like asking which one of your kids would you keep? Each one has a fascination of it's own. Some of the places I've been have grabbed headlines, and some haven't. But for me, I've had backpacking trips just as exciting as the shuttle missions."

Education:

Kathryn Sullivan graduated from Taft High School, Woodland Hills, California, in 1969. She received a Bachelor of Science degree in Earth sciences from the University of California, Santa Cruz, in 1973, and a doctorate in Geology from Dalhousie University (Halifax, Nova Scotia) in 1978.

Experience:

Most of Dr. Sullivan's efforts prior to joining NASA were concentrated in academic study and research. She was an Earth Sciences major at the University of California, Santa Cruz, and spent 1971-1972 as an exchange student at the University of Bergen, Norway. Her bachelor's degree (with honors) was awarded in 1973.

Her doctoral studies at Dalhousie University included participation in a variety of oceanographic expeditions, under the auspices of the U.S. Geological Survey, Wood's Hole Oceanographic Institute, and the Bedford Institute. Her research included the Mid-Atlantic Ridge, the Newfoundland Basin, and fault zones off the Southern California Coast.

Dr. Sullivan is an oceanography officer in the U.S. Naval Reserve, currently holding the rank of lieutenant commander.

Dr. Sullivan left NASA in August 1992 to assume the position of Chief Scientist, National Oceanic and Atmospheric Administration (NOAA). At NOAA, she oversaw research and technology programs in several areas, including climate and global change, marine biodiversity and satellite instrumentation. Currently she is a member of the prestigious Pew Ocean Commission, a national committee tasked with reviewing policies to restore and protect the oceans.

In 1985, Dr. Sullivan became Adjunct Professor of Geology at Rice University, Houston, Texas.

She currently serves as President and CEO of the Center of Science & Industry (COSI) in Columbus, Ohio. COSI Columbus is a science learning centers for kids and adults. COSI creates programs and experiences that make learning science fun through hands-on discovery.

Astronaut Experience:

Selected by NASA in January 1978, Dr. Sullivan became an astronaut in August 1979. Dr. Sullivan's research interests focused on remote sensing. Her Shuttle support assignments included: software development; launch and landing lead chase photographer; Orbiter and cargo test, checkout and launch support at Kennedy Space Center, Florida; extravehicular activity (EVA) and spacesuit support crew for several flights; and capsule communicator (CAPCOM) in Mission Control for numerous Shuttle missions.

She qualified as a systems engineer operator in NASA's WB-57F high-altitude research aircraft in 1978 and has participated in several remote-sensing projects in Alaska. She was a co-investigator on the Shuttle Imaging Radar-B (SIR-B) experiment that she flew on Mission STS-41G.

With the completion of her third mission, Dr. Sullivan had logged over 532 hours in space.

Space Flight Experience:

A veteran of three space flights, Dr. Sullivan was a mission specialist on STS-41G (October 5-13, 1984), STS-31 (April 24-29, 1990), and STS-45 (March 24-April 2, 1992). On her first mission, Dr. Sullivan became the first American woman to perform an EVA.

Her first mission, STS-41G, launched from Kennedy Space Center, Florida, on 5 October 1984, with a crew of seven. During their eight-day mission, the crew deployed the Earth Radiation Budget Satellite, conducted scientific observations of the Earth with the OSTA-3 pallet (including the SIR-B radar, FILE, and MAPS experiments) and large format camera (LFC), conducted a satellite refueling demonstration using hydrazine fuel with the Orbital Refueling System (ORS), and conducted numerous in-cabin experiments as well as activating eight "Getaway Special" canisters. Dr. Sullivan and Commander Leestma also successfully conducted a 3.5 hour Extravehicular Activity (EVA) to demonstrate the feasibility of actual satellite refueling. STS-41G completed 132 orbits of the Earth in 197.5 hours, before landing at Kennedy Space Center, Florida, on 13 October 1984.

"The walk was exciting because it had a certain amount of danger in it," she said referring to the hydrazine fuel she had to manipulate. Hydrazine is highly toxic and can spontaneously combust if subjected to excessive heat.

In April 1990, Dr. Sullivan served on the crew of STS-31, which launched from Kennedy Space Center, Florida, on 24 April 1990. During this five-day mission, crewmembers aboard the Space Shuttle *Discovery* deployed the Hubble Space Telescope, and conducted a variety of mid-deck experiments involving the study of protein crystal growth, polymer membrane processing, and the effects of weightlessness and magnetic fields on an ion arc. They also operated a variety of cameras, including both the IMAX in-cabin and cargo bay cameras, for Earth observations from their record setting altitude of 380 miles. Following 76 orbits of the Earth in 121 hours, STS-31 *Discovery* landed at Edwards Air Force Base, California, on 29 April 1990.

On her last mission, Dr. Sullivan served as Payload Commander on STS-45, the first Spacelab mission dedicated to NASA's Mission to Planet Earth. During this nine-day mission, the crew operated the twelve experiments that constituted the ATLAS-1 (Atmospheric Laboratory for Applications and Science) cargo. ATLAS-1 obtained a vast array of detailed measurements of atmospheric chemical and physical properties, which improved our understanding of our climate and atmosphere. In addition, this was the first time an artificial beam of electrons was used to stimulate a man-made auroral discharge.

Organizations:

Sullivan was appointed to the Chief of Naval Operations Executive Panel in 1988. In March 1985, Dr. Sullivan was appointed by President Reagan to the National Commission on Space. The Commission's report, entitled "Pioneering the Space Frontier," laid out goals for U.S. civilian space activities over the next 25 years. She is also a member of the Geological Society of America, the American Geophysical Union, the American Institute of Aeronautics and Astronautics, the Explorers Club, the Society of Woman Geographers, and the Sierra Club.

Special Honors:

NASA Exceptional Service Medal (1988 & 1991); Ten Outstanding Young People of the World Award, Jaycees International (1987); National Air and Space Museum Trophy, Smithsonian Institution (1985); NASA Space Flight Medal (1984 & 1990); AIAA Haley Space Flight Award (1991); AAS Space Flight Achievement Award (1991).

Astronaut Biography of Anna L. Fisher

Personal Information:

Born 24 August 1949, in New York City, New York, but considers San Pedro, California, to be her hometown.

"I loved reading books, I was an avid reader," Anna recalls about her childhood. "But I hated reading books that had a male lead character. I wanted to read a book that had a female lead character, so that I could identify with them. *So I used to read a lot of biographies about women, Elizabeth Blackwell, Marie Curie, that kind of stuff.*"

Anna's favorite subjects in school were math & science. "It's funny that when I was 12 or 13, I thought maybe I could become a doctor, and be a doctor on a space station or a lunar base, or something like that. But you know, by the time I got into college, that looked like a pipe dream." That didn't stop her from trying, *"Inside myself, I always felt I could do anything I wanted, and I didn't accept barriers imposed by other people."*

Today, Anna is the astronaut doctor she dreamed of becoming. After being an active astronaut for several years Anna took time off to raise her kids. Six years later, she went back to being an astronaut. Anna says that the main reason she came back is, *"I love what I do and I wanted to demonstrate that you can take off time, spend time with your family while your children are young, and come back. There are a lot of people who can fly on the shuttle. There's nobody but me who can be my kid's mom."* But that didn't make it easy. "Coming back was the most difficult thing I've ever done in my life, because by that time, I came back in January of '96, pretty much everyone in my group had flown two to four flights, and most of them were gone, there were only a few people left from my group, and the office was full of new people who I didn't even know."

Anna appreciates that NASA and the Astronaut Office have worked with her on returning to the astronaut corps. "They have been supportive. They realize that this is a stressful job and they have to pay attention to both the technical side and the personal side."

Upon her return to NASA, Anna worked extensively with the Russians on the International Space Station. "I have been very surprised at how you can take two nations that have very different philosophies on how to do things, that were Cold War enemies, and to see that we have learned how to work together. And not only have we learned how to work together, we've become friends."

Anna believes exploration will bring the world together. "Mankind needs to explore, it's essential to our spirit, to our well being."

She hopes that eventually we all will have the chance to go into space. "There will come a time when journalists will go up, and people who want to do other things, like artists. I think those things will open up and will become more available to non-technical, non-science, non-engineering types of people."

Anna enjoys snow and water skiing, jogging, flying, scuba diving, reading, photography, and spending time with her two daughters, Kristin Anne and Kara Lynne.

Her final advice, "Don't be afraid to have dreams because you never know. Those dreams may come true."

Education:

Anna Fisher graduated from San Pedro High School, San Pedro, California, in 1967. She received a bachelor of science in Chemistry and a doctor of Medicine from the University of California, Los Angeles, in 1971 and 1976, respectively. Dr. Fisher completed a one-year internship at Harbor General Hospital in Torrance, California in 1977. She received a master of science in Chemistry from the University of California, Los Angeles, in 1987.

Experience:

After graduating from UCLA in 1971, Dr. Fisher spent a year in graduate school in chemistry at UCLA working in the field of x-ray crystallographic studies of metallocarbonanes. She co-authored 3 publications relating to these studies for the Journal of Inorganic Chemistry. She began medical school at UCLA in 1972 and, following graduation in 1976, began an internship at Harbor General Hospital in Torrance, California. After completing that internship, she specialized in emergency medicine and worked in several hospitals in the Los Angeles area.

Astronaut Experience:

NASA selected Dr. Fisher as an astronaut candidate in January 1978. In August 1979, she completed a 1-year training and evaluation period, making her eligible for assignment as a mission specialist on the Space Shuttle.

Following basic astronaut training, Dr. Fisher's early NASA assignments (pre-STS-1 through STS-4) included the following: Crew representative to support development and testing of the Remote Manipulator System (RMS); Crew representative to support development and testing of payload bay door contingency EVA procedures, the extra-small Extravehicular Mobility Unit (EMU), and contingency on-orbit TPS repair hardware and procedures; Verification of flight software at the Shuttle Avionics Integration Laboratory (SAIL) — she reviewed test requirements and procedures for ascent, on-orbit, and RMS software verification — and served as a crew evaluator for verification and development testing for STS-2, 3 and 4.

For STS-5 through STS-7 Dr. Fisher was assigned as a crew representative to support vehicle integrated testing and payload testing at KSC. In addition, Dr. Fisher supported each Orbital Flight Test (STS 1-4) launch and landing (at either a prime or backup site) as a physician in the rescue helicopters, and provided both medical & operational inputs to the development of rescue procedures. Dr. Fisher was also an on-orbit CAPCOM for the STS-9 mission.

After her first flight, Dr. Fisher was assigned as a mission specialist on STS-61H prior to the *Challenger* accident. Following the accident, she worked as the Deputy of the Mission Development Branch of the Astronaut Office, and as the astronaut office representative for Flight Data File issues. She served as the crew representative on the Crew Procedures Change Board. Dr. Fisher served on the Astronaut Selection Board for the 1987 class of astronauts.

Dr. Fisher also served in the Space Station Support Office where she worked part-time in the Space Station Operations Branch. She was the crew representative supporting space station development in the areas of training, operations concepts, and the health maintenance facility.

Dr. Fisher returned to the Astronaut Office in 1996 after an extended leave of absence to raise her family (1989-1996). When she first returned to the Astronaut Office, she was assigned to the Operations Planning Branch to work on the Operational Flight Data File (procedures) and training issues in support of the International Space Station. She served as the Branch Chief of the Operations Planning Branch from June 1997-June 1998.

Following a reorganization of the Astronaut office, she was assigned as the Deputy for Operations/Training of the Space Station Branch from June 1998-June 1999. She oversaw Astronaut Office inputs to the Space Station Program on issues regarding operations, procedures, and training for the ISS.

Next, she served as Chief of the Space Station Branch of the Astronaut Office with oversight responsibility for 40-50 astronauts and support engineers. In that capacity, she coordinated all astronaut inputs to the Space Station Program Office on issues regarding the design, development, and testing of space station hardware. Additionally, she coordinated all Astronaut Office inputs to Space Station operations, procedures, and training and works with the International Partners to negotiate common design requirements and standards for displays and procedures.

Dr. Fisher is currently assigned to the Shuttle Branch and works technical assignments while awaiting an assignment as either a Space Shuttle crewmember on a Space Station assembly mission or as a crewmember aboard the International Space Station. She hopes to fly in space again soon.

Space Flight Experience:

Dr. Fisher was a mission specialist on STS-51A, which launched from Kennedy Space Center, Florida, on 8 November 1984. She was accompanied by Captain Frederick (Rick) Hauck (spacecraft commander), Captain David M. Walker (pilot), and fellow mission specialists, Dr. Joseph P. Allen, and Commander Dale H. Gardner. This was the second flight of the orbiter *Discovery*. During the mission the crew deployed two satellites, Canada's Anik D-2 (Telesat H) and Hughes' LEASAT-1 (Syncom IV-1), and operated the Radiation Monitoring Equipment (RME) device, and the 3M Company's Diffusive Mixing of Organic Solutions (DMOS) experiment. In the first space salvage mission in history the crew also retrieved for return to earth the Palapa B-2 and Westar VI satellites. STS-51A completed 127 orbits of the Earth before landing at Kennedy Space Center, Florida, on 16 November 1984. With the completion of her first flight, Dr. Fisher has logged a total of 192 hours in space.

Special Honors:

Dr. Fisher was awarded a National Science Foundation Undergraduate Research Fellowship in 1970 and 1971. She graduated from UCLA cum laude and with honors in chemistry. She is a recipient of the NASA Space Flight Medal, Lloyd's of London Silver Medal for Meritorious Salvage Operations, Mother of the Year Award 1984, UCLA Professional Achievement Award, UCLA Medical Professional Achievement Award, and the NASA Exceptional Service Medal in 1999.

Astronaut Biography of Margaret Rhea Seddon

Personal Information:

Born 8 November 1947 in Murfreesboro, Tennessee.

Rhea (pronounced "ray") wanted to be an astronaut since she was 14 and saw Alan Shepard become the first American in space. She believed even then that someday women would be astronauts too.

As a girl Rhea was in the Girl Scouts, on the school newspaper staff, the Science Club, Latin Club, Mathematics Honor Society, and National Honor Society. Her parents supported her even though some girls her age thought she was too much of a "brain" and would never get married.

When she first started school Rhea was intimidated and thought her grades were not good enough to get into medical school. She briefly gave up on becoming a doctor and enrolled in the nursing program at Vanderbilt University. Her desire to be a doctor, and eventually an astronaut, drove her to go back to Berkeley after a year in nursing school. With determination she studied hard and graduated with Honors.

After her internship, she took flying lessons to improve her chances of becoming an astronaut. Rhea worked in a hospital emergency room at night and flew during the day to earn her private pilot's license.

Rhea is married to Former Astronaut Robert L. Gibson of Cooperstown, New York. They have three children.

Education:

Rhea Seddon graduated from Central High School in Murfreesboro, Tennessee, in 1965. She received a Bachelor of Arts degree in physiology from the University of California, Berkeley, in 1970, and a doctorate of medicine from the University of Tennessee College of Medicine in 1973.

Experience:

After medical school, Dr. Seddon completed a surgical internship and 3 years of a general surgery residency in Memphis with a particular interest in nutrition in surgery patients.

Between the period of her internship and residency, she served as an Emergency Department physician at a number of hospitals in Mississippi and Tennessee, and served in this capacity in the Houston area in her spare time. Dr. Seddon has also performed clinical research into the effects of radiation therapy and nutrition in cancer patients.

Dr. Seddon retired from NASA in November 1997. She is now the Assistant Chief Medical Officer of the Vanderbilt Medical Group in Nashville, Tennessee.

Astronaut Experience:

Selected as an astronaut candidate by NASA in January 1978, Dr. Seddon became an astronaut in August 1979.

Her work at NASA has been in a variety of areas, including Orbiter and payload software, Shuttle Avionics Integration Laboratory, Flight Data File, Shuttle medical kit and checklist, launch and landing rescue helicopter physician, support crew member for STS-6, crew equipment, membership on NASA's Aerospace Medical Advisory Committee, Technical Assistant to the Director of Flight Crew Operations, and crew communicator (CAPCOM) in the Mission Control Center.

Dr. Seddon was Assistant to the Director of Flight Crew Operations for Shuttle/Mir Payloads. A three-flight veteran with over 722 hours in space, Dr. Seddon was a mission specialist on STS-51D (1985) and STS-40 (1991), and was the payload commander on STS-58 (1993).

In September 1996, NASA detailed her to Vanderbilt University Medical School in Nashville, Tennessee. She assisted in the preparation of cardiovascular experiments that flew aboard Space Shuttle *Columbia* on the Neurolab Spacelab flight in April 1998.

Space Flight Experience:

STS-51D (*Discovery*), April 12-19, 1985, was launched from and returned to land at the Kennedy Space Center, Florida. The crew deployed ANIK-C for Telesat of Canada, and Syncom IV-3 for the U.S. Navy. A malfunction in the Syncom spacecraft resulted in the first unscheduled EVA (spacewalk), rendezvous and proximity operations for the Space Shuttle in an attempt to activate the satellite using the Remote Manipulator System. The crew conducted several medical experiments, activated two "Getaway Specials," and filmed experiments with toys in space. In completing her first space flight Dr. Seddon logged 168 hours in space in 109 Earth orbits.

STS-40 (*Columbia*) Spacelab Life Sciences (SLS-1), June 5-14, 1991, a dedicated space and life sciences mission was launched from the Kennedy Space Center, Florida, and returned to land at Edwards Air Force Base, California. During the nine-day mission the crew performed experiments, which explored how humans, animals and cells respond to microgravity and re-adapt to Earth's gravity on return. Other payloads included experiments designed to investigate materials science, plant biology and cosmic radiation, and tests of hardware proposed for the Space Station Freedom Health Maintenance Facility. The mission was completed in 146 orbits of the Earth, and logged her an additional 218 hours in space.

STS-58 (*Columbia*), Spacelab Life Sciences-2, flew October 18 to November 1, 1993. Dr. Seddon was the Payload Commander on this life science research mission, which received NASA management recognition as the most successful and efficient Spacelab flown to date. During the fourteen day flight the seven-person crew performed neurovestibular, cardiovascular, cardiopulmonary, metabolic, and musculoskeletal medical experiments on themselves and 48 rats, expanding our knowledge of human and animal physiology both on earth and in space flight. In addition, the crew performed 10 engineering tests aboard the Orbiter *Columbia* and 9 Extended Duration Orbiter Medical Project experiments. The mission was accomplished in 225 orbits of the Earth in over 336 hours.

Astronaut Biography of Shannon W. Lucid

Personal Information:

Born 14 January 1943, in Shanghai, China, but considers Bethany, Oklahoma, to be her hometown.

Shannon's parents were missionaries in China. She and her family traveled together throughout China during and after World War II. She spent the first year of her life in a Japanese prisoner of war camp. The family moved to the United States when Shannon was six-years-old. She remembers, *"My parents always encouraged me to do whatever I wanted to. They always said, 'You can do whatever you set your mind to doing.'"*

Shannon has always enjoyed reading, and as a girl her favorite stories were about American pioneers. "I really enjoyed reading stories about the American west and American explorers and all that. I thought that would be really neat, I thought I would like to be an explorer, to go out and explore the universe." So she learned about rockets and dreamed of exploring space. "And of course when I talked to people about this they thought it would be rather crazy, because that was long before America even had a space program. At the same time I was also reading some science fiction books. They didn't have the word *astronaut*, so I thought I would grow up and be a space scientist and explore the universe, because surely there would be something left to do by the time I grew up. I liked science fiction, not science fantasy. I really liked Isaac Asimov. I even wrote him a letter once and he answered me back. So that's how it all started."

Shannon remembers having lots of fun as a child. "I loved to play pioneers," says Shannon. "We built covered wagons, and we just did all sorts of stuff. I had a great time when I wasn't in school. I loved to do science experiments outside."

"I just hated going to school," she confided, "because I had so much to do, I didn't want to sit there all day. College was fine, but up until then it was just awful. I felt when I graduated from high school I had been released from prison."

As a girl she made her room into a science museum. "I built shelves and I had my museum there. I had my bone collection, I had my bug collection, I had my fossil collection, and I had my other country collection. I had everything. It was really a neat little museum."

As a girl growing up in the shadow of the Cold War, Shannon used to lie awake in bed terrified that Soviet nuclear missiles would one day wipe out everyone she knew. Once in Russia training for her mission, she talked to her cosmonaut crewmembers and found that as children, they too had been scared about American nuclear warheads aimed at them. "It was incredible," she told the Sunday Star Times. "That frightened child never dreamed that one day she would be an astronaut working on a Russian space station."

When Russia and the U.S. started to cooperate, Shannon realized that this was more than good public relations. It was a great opportunity to further space exploration because resources would be combined instead of used in Cold War competition.

Living onboard the *Mir* space station for 188 days, life can get pretty routine. But not boring, she told students at Wellington University. "I never got bored. There were always experiments and maintenance to do. But you get to feel very isolated, cut off from normal things you do here on Earth." However, she says the sense of wonder is always close by, "All you have to do is go to the window and look out."

When asked by the students if she played games on the station, Shannon replied, "Yes, we had laptop computers with chess games and card games. But our favorite game was racing through the space station."

Shannon is married to Michael F. Lucid of Indianapolis, Indiana. They have two daughters and one son. She enjoys flying, camping, hiking, and reading.

"I would like to see us go to Mars," said Shannon. "I can remember thinking with my first daughter as we watched the landing on the moon, that hopefully with my first grandchild we'll watch the landing on Mars. But, I don't think that will happen, and that's sad."

Shannon's advice, "You have to realize that you'll need to keep a balance and maybe you won't be the top person in something. But being the very top isn't the most important. The important thing is to do a good job with what you have and enjoy your family and everything else."

Education:

Shannon Lucid graduated from Bethany High School, Bethany, Oklahoma, in 1960. She received a Bachelor of Science degree in Chemistry from the University of Oklahoma in 1963, and Master of Science and doctor of philosophy degrees in biochemistry from the University of Oklahoma in 1970 and 1973, respectively.

Experience:

Dr. Lucid's experience includes a variety of academic assignments, such as teaching assistant at the University of Oklahoma's Department of Chemistry from 1963 to 1964; senior laboratory technician at the Oklahoma Medical Research Foundation from 1964 to 1966; chemist at Kerr-McGee, Oklahoma City, Oklahoma, 1966 to 1968; graduate assistant at the University of Oklahoma Health Science Center's Department of Biochemistry and Molecular Biology from 1969 to 1973; and research associate with the Oklahoma Medical Research Foundation in Oklahoma City, Oklahoma, from 1974 until her selection to the astronaut candidate training program.

Dr. Lucid is a commercial, instrument, and multi-engine rated pilot.

Astronaut Experience:

Selected by NASA in January 1978, Dr. Lucid became an astronaut in August 1979. She is qualified for assignment as a mission specialist on Space Shuttle flight crews.

Some of her technical assignments have included: the Shuttle Avionics Integration Laboratory (SAIL); the Flight Software Laboratory, in Downey, California, working with the rendezvous and proximity operations group; Astronaut Office interface at Kennedy Space Center, Florida, participating in payload testing, Shuttle testing, and launch countdowns; spacecraft communicator (CAPCOM) in the JSC Mission Control Center during numerous Space Shuttle missions; Chief of Mission Support; Chief of Astronaut Appearances.

In February 2002, Dr. Lucid was selected as NASA's Chief Scientist. She will be stationed at NASA Headquarters, Washington D.C., with responsibility for developing and communicating the agency's science and research objectives to the outside world. "I think my job as chief scientist is sort the umbrella that looks at all of the science to make sure that everybody gets their fair share of the budget. And that there is a balanced science program that NASA is doing."

Space Flight Experience:

A veteran of five space flights, Dr. Lucid has logged 5,354 hours (223 days) in space. She served as a mission specialist on STS-51G (June 17-24, 1985), STS-34 (October 18-23, 1989), STS-43 (August 2-11, 1991), STS-58 (October 18 to November 1, 1993), and most recently served as a Board Engineer 2 on Russia's Space Station *Mir* (launching March

22, 1996 aboard STS-76 and returning September 26, 1996 aboard STS-79). Until June 2002, Dr. Lucid held an international record for the most flight hours in orbit by any non-Russian, and still holds the record for the most flight hours in orbit by any woman in the world.

STS-51G *Discovery* (June 17-24, 1985) was a 7-day mission during which crew deployed communications satellites for Mexico (Morelos), the Arab League (Arabsat), and the United States (AT&T Telstar). They used the Remote Manipulator System (RMS) to deploy and later retrieve the SPARTAN satellite, which performed 17 hours of x-ray astronomy experiments while separated from the Space Shuttle. In addition, the crew activated the Automated Directional Solidification Furnace (ADSF), six Getaway Specials, and participated in biomedical experiments. The mission was accomplished in 112 orbits of the Earth, traveling 2.5 million miles in 169 hours and 39 minutes. Landing was at Edwards Air Force Base (EAFB), California.

STS-34 *Atlantis* (October 18-23, 1989) was a 5-day mission during which the crew deployed the Galileo spacecraft on its journey to explore Jupiter, operated the Shuttle Solar Backscatter Ultraviolet Instrument (SSBUV) to map atmospheric ozone, and performed numerous secondary experiments involving radiation measurements, polymer morphology, lightning research, microgravity effects on plants, and a student experiment on ice crystal growth in space. The mission was accomplished in 79 orbits of the Earth, traveling 1.8 million miles in 119 hours and 41 minutes. Landing was at Edwards Air Force Base, California.

STS-43 *Atlantis* (August 2-11, 1991) was a nine-day mission during which the crew deployed the fifth Tracking and Data Relay Satellite (TDRS-E). The crew also conducted 32 physical, material, and life science experiments, mostly relating to the Extended Duration Orbiter and Space Station Freedom. The mission was accomplished in 142 orbits of the Earth, traveling 3.7 million miles in 213 hours, 21 minutes, 25 seconds. STS-43 *Atlantis* was the eighth Space Shuttle to land at KSC).

STS-58 *Columbia* (October 18 to November 1, 1993). This record duration fourteen-day mission was recognized by NASA management as the most successful and efficient Spacelab flight flown by NASA. The STS-58 crew performed neurovestibular, cardiovascular, cardiopulmonary, metabolic, and musculoskeletal medical experiments on themselves and 48 rats, expanding our knowledge of human and animal physiology both on Earth and in space flight. In addition, they performed 16 engineering tests aboard the *Columbia* and 20 Extended Duration Orbiter Medical Project experiments. The mission was accomplished in 225 orbits of the Earth, traveling 5.8 million miles in 336 hours, 13 minutes, and 1 second. Landing was at Edwards Air Force Base, California. In completing this flight Dr. Lucid logged 838 hours, 54 minutes in space making her the American woman astronaut with the most hours in space.

Dr. Lucid currently holds the United States single mission space flight endurance record on the Russian Space Station Mir. Following a year of training in Star City, Russia, her journey started with liftoff at Kennedy Space Center, Florida, on March 22, 1996 aboard STS-76 *Atlantis*. Following docking, she transferred to the *Mir* Space Station. Assigned as a Board Engineer 2, she performed numerous life science and physical science experiments during the course of her stay aboard Mir. Her return journey to KSC was made aboard STS-79 *Atlantis* on September 26, 1996. In completing this mission Dr. Lucid traveled 75.2 million miles in 188 days, 04 hours, 00 minutes, 14 seconds.

Special Honors:

Dr. Lucid is the recipient of numerous awards. President Bill Clinton awarded Dr. Lucid the Congressional Space Medal of Honor in 1996. She is the first and only woman to have earned this prestigious award.

Dr. Lucid was also awarded the Order of Friendship Medal by Russian President Boris Yeltsin. This is one of the highest Russian civilian awards and the highest award that can be presented to a non-citizen.

Astronaut Biography of Bonnie Dunbar

Personal Information:

Born 3 March 1949, in Sunnyside, Washington.

Bonnie grew up on a farm her parents homesteaded in Washington State. The nearest town was several miles away, which meant that she had few playmates, but great star watching at night, which she did frequently. She read a lot and her favorite books were the classics and science fiction.

"I wanted to be an astronaut at a time when there weren't any women astronauts," Bonnie recalls. "When I was growing up. Women couldn't do what I wanted to do. Yet I clung to my goal. So I looked toward the future. I knew that someday women would become astronauts, and when they did, I wanted to be as qualified for the job as possible. I loved aviation as a teenager, so I thought that I'd become an astronaut by starting out as a military test pilot. But back then, women weren't admitted into Air Force or Navy programs. I applied later, when they started admitting women, but officials told me I was too old. So I had to take a different route."

When her initial plans didn't work out, Bonnie kept going with energy. She doesn't "agonize over the things that I can't change. *There's no better way to waste time in life than to dwell on the past. You should learn from the past with as much objectivity as possible, so that you can modify your future actions - and then move on."*

Soon the astronaut program was looking for people with Bonnie's technical expertise. "I studied engineering and eventually became a mission specialist. I still get to fly, and I have a rewarding and productive job, overseeing scientific experiments and other operations that take place in space."

So her lesson is to remain optimistic and do what you can. It worked out for her. *"I always believe that if you remain optimistic, and if you prepare yourself for opportunities, then those opportunities will find you. I tell young women not to let themselves feel discouraged just because there's a fence on this side of the pasture: the gate may be open on the other side."*

Education:

Bonnie Dunbar graduated from Sunnyside High School, Sunnyside, Washington, in 1967. She received a Bachelor of Science and Master of Science degrees in ceramic engineering from the University of Washington in 1971 and 1975, respectively. Her doctorate in Mechanical/Biomedical Engineering from the University of Houston was awarded in 1983.

Experience:

Following graduation in 1971, Dr. Dunbar worked for Boeing Computer Services for two years as a systems analyst. From 1973 to 1975, she conducted research for her master's thesis in the field of mechanisms and kinetics of ionic diffusion in sodium beta-alumina. In 1975, she was invited to participate in research at Harwell Laboratories in Oxford, England, as a visiting scientist. Her work there involved the wetting behavior of liquids on solid substrates. Following her work in England, she accepted a senior research engineer position with Rockwell International Space Division in Downey, California. Her responsibilities there included developing equipment and processes for the manufacture of the Space Shuttle thermal protection system in Palmdale, California. She also represented Rockwell International as a member of the Dr. Kraft Ehricke evaluation committee on prospective space industrialization concepts.

Dr. Dunbar completed her doctorate at the University of Houston in Houston, Texas. Her multi-disciplinary dissertation (materials science and physiology) involved evaluating the effects of simulated space flight on bone strength and fracture toughness. These results were correlated to alterations in hormonal and metabolic activity. Dr. Dunbar has served as an adjunct assistant professor in Mechanical Engineering at the University of Houston.

She is a private pilot with over 200 hours in single engine land aircraft, has logged more than 700 hours flying time in T-38 jets as co-pilot, and has over 100 hours as co-pilot in a Cessna Citation Jet.

Astronaut Experience:

Dr. Dunbar accepted a position as a payload officer/flight controller at the Lyndon B. Johnson Space Center in 1978. She served as a guidance and navigation officer/flight controller for the *Skylab* re-entry mission in 1979 and was subsequently designated project officer/payload officer for the integration of several Space Shuttle payloads.

Dr. Dunbar became a NASA astronaut in August 1981. Her technical assignments have included assisting in the verification of Shuttle flight software at the Shuttle Avionics Integration Laboratory (SAIL), serving as a member of the Flight Crew Equipment Control Board, participation as a member of the Astronaut Office Science Support Group, supporting operational development of the remote manipulator system (RMS). She has served as chief of the Mission Development Branch, as the Astronaut Office interface for "secondary" payloads, and as lead for the Science Support Group. In 1993, Dr. Dunbar served as Deputy Associate Administrator, Office of Life and Microgravity Sciences, NASA Headquarters, Washington, D.C.

In February 1994, she traveled to Star City, Russia, where she spent 13-months training as a back-up crewmember for a 3-month flight on the Russian Space Station, Mir. In March 1995, she was certified by the Russian Gagarin Cosmonaut Training Center as qualified to fly on long duration *Mir* Space Station flights. From October 1995 to November 1996, she was detailed to the NASA JSC Mission Operations Directorate as Assistant Director where she was responsible for chairing the International Space Station Training Readiness Reviews, and facilitating Russian/American operations and training strategies.

Currently, Dr. Dunbar serves as Assistant Director to the NASA Johnson Space Center (JSC) with a focus on University Research.

Space Flight Experience:

A veteran of five space flights, Dr. Dunbar has logged more than 1,208 hours (50 days) in space. She served as a mission specialist on STS 61-A in 1985, STS-32 in 1990, and STS-71 in 1995, and was the Payload Commander on STS-50 in 1992, and STS-89 in 1998.

STS 61-A *Challenger* (October 30-November 6, 1985) was the West German D-1 Spacelab mission. It was the first to carry eight crewmembers, the largest to fly in space, and was also the first in which payload activities were controlled from outside the United States. More than 75 scientific experiments were completed in the areas of physiological sciences, materials science, biology, and navigation. During the flight, Dr. Dunbar was responsible for operating Spacelab and its subsystems and performing a variety of experiments. Her mission training included six months of experiment training in Germany, France, Switzerland, and The Netherlands. STS 61-A launched from the Kennedy Space Center, Florida, and returned to land at Edwards Air Force Base, California. Mission duration was 7 days, 44 minutes 51 seconds, traveling 2.5 million miles in 111 orbits of the Earth.

STS-32 *Columbia* (January 9-20, 1990), launched from the Kennedy Space Center, Florida, and returned to a night landing at Edwards Air Base in California. During the flight, the crew successfully deployed the Syncom IV-F5 satellite, and retrieved the 21,400-pound Long Duration Exposure Facility (LDEF) using the RMS. They also operated a variety of mid-deck experiments including the Microgravity Disturbance Experiment (MDE) using the Fluids Experiment Apparatus (FEA), Protein Crystal Growth (PCG), American Flight Echocardiograph (AFE), Latitude/Longitude Locator (L3), Mesoscale Lightning Experiment (MLE), Characterization of Neurospora Circadian Rhythms (CNCR), and the IMAX Camera. Dr. Dunbar was principal investigator for the MDE/FEA Experiment. Additionally, numerous medical test objectives, including in-flight lower body negative pressure (LBNP), in-flight aerobic exercise and muscle performance were conducted to evaluate human adaptation to extended duration missions. Mission duration was 10 days, 21 hours, 01 minute, 38 seconds, traveling 4.5 million miles in 173 orbits of the Earth.

STS-50 *Columbia* (June 25 to July 9, 1992). Dr. Dunbar was the Payload Commander on STS-50, the United States Microgravity Lab-1 mission that was dedicated to microgravity fluid physics and materials science. Over 30 experiments sponsored by over 100 investigators were housed in the "Spacelab" in the Shuttle's Payload Bay. A payload crew of four operated around-the-clock for 13 days performing experiments in scientific disciplines such as protein crystal growth, electronic and infrared detector crystal growth, surface tension physics, zeolite crystal growth, and human physiology. Mission duration was 13 days, 19 hours, 30 minutes and 4 seconds, traveling 5.7 million miles in 221 orbits of the Earth. STS-71 *Atlantis* (June 27 to July 7, 1995), was the first Space Shuttle mission to dock with the Russian Space Station Mir, and involved an exchange of crews. The *Atlantis* was modified to carry a docking system compatible with the Russian *Mir* Space Station. Dr. Dunbar served as MS-3 on this flight, which also carried a Spacelab module in the payload bay in which the crew performed medical evaluations on the returning *Mir* crew. These evaluations included ascertaining the effects of weightlessness on the cardio/vascular system, the bone/muscle system, the immune system, and the cardio/pulmonary system. Mission duration was 9 days, 19 hours, 23 minutes and 8 seconds, traveling 4.1 million miles in 153 orbits of the earth.

STS-89 Endeavour (January 22-31, 1998), was the eighth Shuttle-Mir docking mission during which the crew transferred more than 9,000 pounds of scientific equipment, logistical hardware and water from Space Shuttle Endeavour to Mir. In the fifth and last exchange of a U.S. astronaut, STS-89 delivered Andy Thomas to *Mir* and returned with David Wolf. Mission duration was 8 days, 19 hours and 47 seconds, traveling 3.6 million miles in 138 orbits of the Earth. Dr. Dunbar was the Payload Commander, responsible for all payload activities including the conduct of 23 technology and science experiments.

Organizations:

Dr. Dunbar is a member of the American Ceramic Society (ACS), the National Institute of Ceramic Engineers (NICE), the Society of Biomedical Engineering, American Association for the Advancement of Science, Materials Research Society (MRS), and the National Science Foundation (NSF) Engineering Advisory Board, 1993 to present.

She is also on the Board of Directors, Arnold Air Society and Angel Flight, International Academy of Astronautics (IAF), Experimental Aircraft Association (EAA), Society of Women Engineers (SWE), Association of Space Explorers (ASE).

Special Honors:

Dr. Dunbar has received the American Ceramics Society James I. Mueller Award in Cocoa Beach, Florida. (2000). She was inducted into the Women in Technology International (WITI) Hall of Fame in 2000, one of five women in the world so honored and was selected as one of the top 20 women in technology in Houston, Texas (2000). Some of Dr. Dunbar awards include five NASA Space Flight Medals (1985, 1990, 1992, 1995 and 1998), a Superior Accomplishment Award (1997), NASA Exceptional Achievement Medal (1996), NASA Outstanding Leadership Award (1993), Design News Engineering Achievement Award (1993), IEEE Judith Resnik Award (1993), Society of Women Engineers Resnik *Challenger* Medal (1993), Boeing Corporation Pathfinder Award (1992). AAES National Engineering Award (1992), and a NASA Exceptional Service Award (1991).

Astronaut Biography of Mary L. Cleave

Personal Information:

Born 5 February 1947, in Southampton, New York.

At a Philosophical Society of Washington meeting, Mary told the group, "Weighing nothing is great fun, especially if you're short. You don't need a stool to reach things!"

But the view from space is what Mary comments on most. "Sunrise, sunset every 45 minutes — they're gorgeous. But you also get an appreciation for just how small the atmosphere is. It's just a very thin layer that separates us and our planet from a really nasty vacuum out there. So you get real respectful for how we should treat the atmosphere, how we should treat the oceans, how we should treat the land."

Mary says most astronauts are not trained as environmentalists, who are taught to look at conservation and preservation of the planet. But being in space changes a person's attitude. "This conversion usually comes about after the initial and most striking observations from space that our planet is quite finite and it's also 70% ocean. Whereas, we know these things intellectually from our Earth-bound field trips and science books, the finite nature of our planet and the small portion that we actually inhabit instantly becomes real when you look down from low Earth orbit."

Education:

Mary Cleave graduated from Great Neck North High School, Great Neck, New York, in 1965. She received a Bachelor of Science degree in Biological Sciences from Colorado State University in 1969 and Master of Science in Microbial Ecology and a doctorate in Civil and Environmental Engineering from Utah State University in 1975 and 1979, respectively.

Experience:

Dr. Cleave held graduate research and research engineer assignments in the Ecology Center and the Utah Water Research Laboratory at Utah State University from September 1971 to June 1980.

Her work included research on a number of microbial, environmental, and engineering projects including:
The productivity of the algal component of cold desert soil crusts in the Great Basin Desert south of Snowville, Utah. Algal removal with intermittent sand filtration. Prediction of minimum river flow necessary to maintain certain game fish. The effects of increased salinity and oil shale leachates on freshwater phytoplankton productivity. Development of the Surface Impoundment Assessment for current and future processing of data from surface impoundments in Utah. Design and implementation of an algal bioassay center and a workshop for bioassay techniques for the Intermountain West.

In conjunction with her research efforts, she has published numerous scientific papers.

Dr. Cleave was selected as an astronaut in May 1980. Her technical assignments have included: flight software verification in the Shuttle Avionics Integration Laboratory (SAIL); CAPCOM on five Space Shuttle flights; Malfunctions Procedures Book; Crew Equipment Design.

Dr. Cleave left JSC in May 1991 to join NASA's Goddard Space Flight Center (GSFC), where she joined the Laboratory for Hydrospheric Processes in 1991. She was the Project Manager for the Sea-viewing Wide Field-of-view Sensor (SeaWiFS), an ocean color sensor that is monitoring global marine chlorophyll. SeaWiFS launched on August 1, 1997.

After the SeaWiFS Project became operational, she entered the NASA Senior Executive Service Career Development Program. She was then the Earth Science Representative to NASA's Chief Scientist

Dr. Mary Cleave was appointed as Deputy Associate Administrator of NASA's Office of Earth Science in February 2000. Her primary responsibility is developing plans and priorities for the Earth Science Enterprise's advanced science, technology, and applications.

Space Flight Experience:

A veteran of two space flights, Dr. Cleave has logged a total of 10 days, 22 hours, 02 minutes, 24 seconds in space, orbited the earth 172 times and traveled 3.94 million miles. She was a mission specialist on STS 61-B (November 26 to December 3, 1985) and STS-30 (May 4-8, 1989).

STS-61B *Atlantis* (Nov. 26 to Dec. 3, 1985) launched at night from the Kennedy Space Center, Florida, and returned to land on Runway 22 at Edwards Air Force Base, California. During the mission, the crew deployed the MORELOS-B, AUSSAT II, and SATCOM K-2 communications satellites, conducted 2 six-hour spacewalks to demonstrate space station construction techniques with the EASE/ACCESS experiments, operated the Continuous Flow Electrophoresis (CFES) experiment for McDonnell Douglas and a Getaway Special (GAS) container for Telesat, Canada, conducted several Mexican Payload Specialist Experiments for the Mexican Government, and tested the Orbiter Experiments Digital Autopilot (OEX DAP). This was the heaviest payload weight carried to orbit by the Space Shuttle to date. Mission duration was 165 hours, 4 minutes, 49 seconds.

STS-30 *Atlantis* (May 4-8, 1989) was a four day mission during which the crew successfully deployed the Magellan Venus-exploration spacecraft, the first U.S. planetary science mission launched since 1978, and the first planetary probe to be deployed from the Shuttle. Magellan arrived at Venus in August 1990 and mapped over 95% of the surface of Venus. Magellan has been one of NASA's most successful scientific missions providing valuable information about the Venusian atmosphere and magnetic field. In addition, the crew also worked on secondary payloads involving Indium crystal growth, electrical storm, and earth observation studies. Mission duration was 96 hours, 57 minutes, 35 seconds.

Organizations:

Dr. Cleave is a member of the Texas Society of Professional Engineers, the Water Pollution Control Federation, Tri-Beta, Sigma XI, and Tau Beta Pi. She is also an association member of the American Society of Civil Engineers.

Astronaut Biography of Sharon Christa Corrigan McAuliffe

Personal Information:

Born 2 September 1948 in Boston, Massachusetts.

Christa loved teaching. She believed that hands on learning was the best kind, and was often called the field trip teacher. Her students say she was an "inspirational human being, a marvelous teacher who made their lessons come alive." Before her launch, Christa called her upcoming trip on the *Challenger* "The Ultimate Field Trip."

At a press conference on 6 December 1985, Christa said, "Space is for everybody. It's not just for a few people in science or math, or for a select group of astronauts. That's our new frontier out there, and it's everybody's business to know about space."

Christa's mother, Grace Corrigan, said in her book *A Journal For Christa*, "Christa lived. She never just sat back and existed. Christa always accomplished everything that she was capable of accomplishing. She extended her own limitations. She cared about her fellow human beings. She did the ordinary, but she did it well and unfailingly."

Christa's recreational interests included jogging, tennis, and volleyball.

In her winning essay to NASA to become the first teacher in space, Christa said, "I cannot join the space program and restart my life as an astronaut, but this opportunity to connect my abilities as an educator with my interests in history and space is a unique opportunity to fulfill my early fantasies."

Education:

Christa McAuliffe graduated from Marian High School, Framingham, Massachusetts, in 1966. She received a Bachelor of Arts degree, Framingham State College in 1970 and a Masters degree in education, Bowie State College, Bowie, Maryland in 1978.

Experience:

From 1970-1971 McAuliffe taught at Benjamine Foulois Junior High School, Morningside, Maryland.

From 1971-1978 she taught 8th grade American history at Thomas Johnson Junior High School, Lanham, Maryland.

From 1978-1979 she taught 8th grade English and American history, and 9th grade Civics at Bundlett Junior High School, Concord, New Hampshire.

From 1980-1982 she taught 7th and 8th grade American history at Bow Memorial High School, Concord, New Hampshire. 1982-1985 she taught 9th grade English at Concord High School, Concord, New Hampshire.

McAuliffe also taught courses in economics, law, American history, and a course she developed entitled "The American Woman," to students in the 10th, 11th, and 12th grade.

Astronaut Experience:

Christa McAuliffe was selected as the primary candidate for the NASA Teacher in Space Project on 19 July 1985. She was a payload specialist on STS 51-L, which was launched from the Kennedy Space Center, Florida, at 11:38:00 EST on 28 January 1986. The crew on board the *Challenger* included the spacecraft commander, Mr. F.R. Scobee, the pilot, Commander M.J. Smith (USN), mission specialists, Dr. R.E. McNair, and Lieutenant Colonel E.S. Onizuka (USAF), and fellow civilian payload specialists G.B. Jarvis and J. Resnik. The STS 51-L crew died when the *Challenger* exploded 72 seconds after launch.

Organizations:

Christa McAuliffe was a Board member for the New Hampshire Council of Social Studies. She was also a member of the National Council of Social Studies, Concord Teachers Association, New Hampshire Education Association, and the National Education Association.

Astronaut Biography of Ellen S. Baker

Personal Information:

Born 27 April 1953, in Fayetteville, North Carolina, but considers New York City her hometown.

As a girl Ellen had big dreams, "I wanted to play baseball. But they didn't let girls play major league baseball, so it was a big blow."

Fortunately baseball wasn't her only interest. "I've always been interested in the space program and I thought it was exciting and challenging," explains Ellen. "But they didn't let girls go into space when I was little. So I didn't really think about it as a possibility for me until I was out of medical school. The first group of women were selected in 1978 and that's the year I graduated from medical school."

Now that she's an astronaut, Ellen loves it. "What I like best are the people. I work with a wonderful group of people and it's hard to imagine that there's a better group of people anywhere. And of course I love the program, the excitement, the challenge, and I love working for something that I think is important. I love to make a difference."

Ellen is married to Kenneth J. Baker. They have two daughters. Ellen enjoys swimming, skiing, running, movies, music, and reading. *Ellen's advice is, "work really hard and do your very best at whatever it is you choose to do. Find something you really like to do and do it the very best you can."*

Education:

Ellen Baker graduated from Bayside High School, New York, New York, in 1970. She received a Bachelor of Arts degree in geology from the State University of New York at Buffalo in 1974, a doctorate of medicine degree from Cornell University in 1978, and a Masters in public health from University of Texas School of Public Health in 1994.

Experience:

After completing medical school, Dr. Baker trained in internal medicine at the University of Texas Health Science

Center, San Antonio, Texas. In 1981, after three years of training, she was certified by the American Board of Internal Medicine.

Astronaut Experience:

In 1981, following her residency, Dr. Baker joined NASA as a medical officer at the Johnson Space Center. That same year, she graduated from the Air Force Aerospace Medicine Course at Brooks Air Force Base, San Antonio, Texas. Prior to her selection as an astronaut candidate she served as a physician in the Flight Medicine Clinic at the Johnson Space Center.

Selected by NASA in May 1984, Dr. Baker became an astronaut in June 1985. Since then, she has worked at a variety of jobs at NASA in support of the space shuttle program and Space Station development.

Space Flight Experience:

A veteran of three space flights, Dr. Baker has logged over 686 hours in space. She was a mission specialist on STS-34 in 1989, STS-50 in 1992, and STS-71 in 1995.

STS-34 *Atlantis* (October 18-23, 1989) launched from the Kennedy Space Center in Florida, and landed at Edwards Air Force Base in California. During the mission, the crew successfully deployed the Galileo probe to explore Jupiter, operated the Shuttle Solar Backscatter Ultraviolet Instrument (SSBUV) to map atmospheric ozone, conducted several medical experiments, and numerous scientific experiments. Mission objectives were accomplished in 79 orbits of the Earth, traveling 1.8 million miles in 119 hours and 41 minutes.

STS-50 *Columbia* (June 25-July 9, 1992) launched and landed at the Kennedy Space Center in Florida. STS-50 was the first flight of the United States Microgravity Laboratory and the first Extended Duration Orbiter flight. Over a two-week period, the crew conducted scientific experiments involving crystal growth, fluid physics, fluid dynamics, biological science, and combustion science. Mission objectives were accomplished in 221 orbits of the Earth, traveling 5.7 million miles in 331 hours 30 seconds and 4 minutes in space.

STS-71 *Atlantis* (June 27-July 7, 1995) launched from the Kennedy Space Center with a seven-member crew and returned there with an eight-member crew. STS-71 was the first Space Shuttle mission to dock with the Russian Space Station Mir, and involved an exchange of crews. The *Atlantis* Space Shuttle was modified to carry a docking system compatible with the Russian *Mir* Space Station. It also carried a Spacelab module in the payload bay in which the crew performed various life sciences experiments and data collections. Mission accomplished in 153 orbits of the Earth, traveling 4.1 million miles in 235 hours and 23 minutes.

Astronaut Biography of Kathryn C. Thornton

Personal Information:

Born 17 August 1952 in Montgomery, Alabama.

Kathy dreamed of becoming an astronaut since she was a girl. She remembers her brothers and sisters riding skateboards in the driveway and talking to each other on walkie-talkies. Kathy says they took turns as "Mission Control" on pretend space missions and she loved to be the astronaut.

Before she was an astronaut, she became a physicist and went into the Army. While stationed and working at the Army's Science and Technology Center (Charlottesville, Virginia), Thornton saw a NASA announcement looking for space shuttle astronauts. She told *Southern Living Magazine,* "The interviews were hard, but once you got into the program, the training was even more intense."

Kathy is married to Stephen T. Thornton, Ph.D., of Oak Ridge, Tennessee. She has two stepsons and three daughters. She enjoys scuba diving and skiing.

Education:

Kathryn Thornton graduated from Sidney Lanier High School, Montgomery, Alabama, in 1970. She received a Bachelor of Science degree in physics from Auburn University in 1974, a Master of Science degree in physics from the University of Virginia in 1977, and a doctorate of philosophy in physics from the University of Virginia in 1979.

Experience:

After Dr. Thornton earned her Ph.D. at the University of Virginia in 1979, she was awarded a NATO Postdoctoral Fellowship to continue her research at the Max Planck Institute for Nuclear Physics in Heidelberg, West Germany. In 1980, she returned to Charlottesville, Virginia, where she was employed as a physicist at the U.S. Army Foreign Science and Technology Center.

Dr. Thornton left NASA in August 1996, to join the faculty of the University of Virginia.

Astronaut Experience:

Selected by NASA in May 1984, Dr. Thornton became an astronaut in July 1985. Her technical assignments have included flight software verification in the Shuttle Avionics Integration Laboratory (SAIL), serving as a team member of the Vehicle Integration Test Team (VITT) at KSC, and as a spacecraft communicator (CAPCOM).

Space Flight Experience:

A veteran of three space flights, Dr. Thornton flew on STS-33 in 1989, STS-49 in 1992, and STS-61 in 1993. She has logged over 975 hours in space, including more than 21 hours of extravehicular activity (EVA).

Dr. Thornton was a mission specialist on the crew of STS-33, which launched at night from Kennedy Space Center, Florida, on November 22, 1989, aboard the Space Shuttle *Discovery*. The mission carried Department of Defense payloads and other secondary payloads. After 79 orbits of the Earth, this five-day mission concluded on November 27, 1989, at Edwards Air Force Base, California.

On her second flight, Dr. Thornton served on the crew of STS-49, May 7-16, 1992, on board the maiden flight of the new Space Shuttle Endeavour. During the mission the crew conducted the initial test flight of the *Endeavour*, performed a record four EVA's (space walks) to retrieve, repair and deploy the International Telecommunications Satellite (INTELSAT), and to demonstrate and evaluate numerous EVA tasks to be used for the assembly of Space Station *Freedom*.

Dr. Thornton was one of two EVA crewmembers that evaluated Space Station assembly techniques on the fourth EVA. STS-49 logged 213 hours in space and 141 Earth orbits prior to landing at Edwards Air Force Base, California.

On her third flight, Dr. Thornton was a mission specialist EVA crew member aboard the Space Shuttle Endeavour on the STS-61 Hubble Space Telescope (HST) servicing and repair mission. STS-61 launched at night from the Kennedy Space Center, Florida, on December 2, 1993. During the 11-day flight, the HST was captured and restored to full capacity through a record five space walks by four astronauts. After having traveled 4,433,772 miles in 163 orbits of the Earth, the crew of Endeavour returned to a night landing at the Kennedy Space Center on December 13, 1993.

From October 20 to November 5, 1995, Dr. Thornton served aboard Space Shuttle *Columbia* on STS-73, as the payload commander of the second United States Microgravity Laboratory mission. The mission focused on materials science, biotechnology, combustion science, the physics of fluids, and numerous scientific experiments housed in the pressurized Spacelab module. In completing her fourth space flight, Dr. Thornton orbited the Earth 256 times, traveled over 6 million miles, and logged a total of 15 days, 21 hours, 52 minutes and 21 seconds in space.

Organizations:

Dr. Thornton is a member of the American Physical Society, American Association for the Advancement of Science, Sigma Xi, Phi Kappa Phi, and Sigma Pi Sigma.

Astronaut Biography of Marsha S. Ivins

Personal Information:

Born 15 April 1951, in Baltimore, Maryland.

Marsha enjoys music by the Grammy award-winning banjo player Alison Brown and her band, the Alison Brown Quartet. Apparently Ms. Brown received a rare fan letter from the astronaut that inspired her to write the song, "My Favorite Marsha."

Education:

Graduated from Nether Providence High School, Wallingford, Pennsylvania, in 1969; received a bachelor of science degree in aerospace engineering from the University of Colorado in 1973.

Astronaut Experience:

Ms. Ivins has worked at the Johnson Space Center since July 1974, and until 1980, was assigned as an engineer, working on orbiter displays and controls and man machine engineering. Her major assignment in 1978 was to participate in development of the Orbiter Heads-Up Display (HUD). In 1980 she was assigned as a flight engineer on the Shuttle Training Aircraft (Aircraft Operations) and a co-pilot in the NASA administrative aircraft (Gulfstream-1).

Ms. Ivins holds a multi-engine Airline Transport Pilot License with Gulfstream-1 type rating, single engine airplane, land, sea, and glider commercial licenses, and airplane, instrument, and glider flight instructor ratings. She has logged over 6,000 hours in civilian and NASA aircraft.

Ms. Ivins was selected in the NASA Astronaut Class of 1984 as a Mission Specialist. Her technical assignments to date include: review of Orbiter safety and reliability issues; avionics upgrades to the Orbiter cockpit; software verification in the Shuttle Avionics Integration Laboratory (SAIL); Spacecraft Communicator (CAPCOM) in Mission Control; crew representative for Orbiter photographic system and procedures; crew representative for Orbiter flight crew equipment issues; Lead of Astronaut Support Personnel team at the Kennedy Space Center in Florida, supporting space shuttle launches and landings; crew representative for Space Station stowage, habitability, logistics, and transfer issues.

Space Flight Experience:

A veteran of five space flights, (STS-32 in 1990, STS-46 in 1992, STS-62 in 1994, STS-81 in 1997, and STS-98 in 2001); Ms. Ivins has logged over 1,318 hours in space.

STS-32 (January 9-20, 1990) launched from the Kennedy Space Center, Florida, on an eleven-day flight, during which crew members onboard the orbiter *Columbia* successfully deployed a Syncom satellite, and retrieved the 21,400-pound Long Duration Exposure Facility (LDEF). Mission duration was 261 hours, 1 minute, and 38 seconds. Following 173 orbits of the Earth and 4.5 million miles, *Columbia* returned with a night landing at Edwards Air Force Base, California.

STS-46 (July 31-August 8, 1992) was an 8-day mission, during which crew-members deployed the European Retrievable Carrier (EURECA) satellite, and conducted the first Tethered Satellite System (TSS) test flight. Mission duration was 191 hours, 16 minutes, and 7 seconds. Space Shuttle *Atlantis* and her crew launched and landed at the Kennedy Space Center, Florida, completing 126 orbits of the Earth in 3.35 million miles.

STS-62 (March 4-18, 1994) was a 14-day mission for the United States Microgravity Payload (USMP) 2 and Office of Aeronautics and Space Technology (OAST) 2 payloads. These payloads studied the effects of microgravity on materials sciences and other space flight technologies. Other experiments on board included demonstration of advanced teleoperator tasks using the remote manipulator system, protein crystal growth, and dynamic behavior of space structures. Mission duration was 312 hours, 23 minutes, and 16 seconds. Space Shuttle *Columbia* launched and landed at the Kennedy Space Center, Florida, completing 224 orbits in 5.82 million miles.

STS-81 *Atlantis* (January 12-22, 1997) was a 10-day mission, the fifth to dock with Russia's Space Station Mir, and the second to exchange U.S. astronauts. The mission also carried the Spacehab double module providing additional mid-deck locker space for secondary experiments. In five days of docked operations more than three tons of food, water, experiment equipment and samples were moved back and forth between the two spacecraft. Following 160 orbits of the Earth the STS-81 mission concluded with a landing on Kennedy Space Center's Runway 33 ending a 3.9 million mile journey. Mission duration was 244 hours, 56 minutes.

STS-98 *Atlantis* (February 7-20, 2001) continued the task of building and enhancing the International Space Station by delivering the U.S. laboratory module Destiny. The Shuttle spent seven days docked to the station while Destiny was attached and three spacewalks were conducted to complete its assembly. The crew also relocated a docking port, and delivered supplies and equipment to the resident Expedition-1 crew. Space Shuttle *Atlantis* returned to land at Edwards Air Force Base, California. Mission duration was 12 days, 21 hours, 20 minutes.

Astronaut Biography of Linda M. Godwin

Personal Information:

Born 2 July 1952, in Cape Girardeau, Missouri.

Linda grew up with an interest in math and science. *"I had some good math and science teachers, you know, going through school, and somehow instead of turning me off, they made it seem interesting. And my high school science teachers have come to all my shuttle launches so far."*

It wasn't until after she was finished with college that Linda thought about being an astronaut. "I grew up watching a lot of the coverage of the early U.S. space program, all the way back, starting with *Mercury* and then through *Gemini* and *Apollo* and of course to the moon as the main part of the *Apollo* program. So that made me interested in NASA, but I never thought it was something I could do. But I imagined that it fueled an interest in the science that I had already. So I chose to go into physics when I went to college."

As a result, Linda was ready when NASA began selecting astronauts again. "I was fortunate enough to have chosen some educational routes that put me in a position where I could at least have a shot at it."

Linda is married to Steven R. Nagel of Houston, Texas. They have two daughters.

Linda is an instrument rated private pilot.

Education:

Linda Godwin graduated from Jackson High School in Jackson, Missouri, in 1970. She received a Bachelor of Science degree in mathematics and physics from Southeast Missouri State in 1974, and a Master of Science degree and a doctorate in physics from the University of Missouri in 1976 and 1980.

Experience:

While at graduate school Linda Godwin taught undergraduate physics labs and was the recipient of several research assistantships. She conducted research in low temperature solid-state physics, including studies in electron tunneling and vibrational modes of absorbed molecular species on metallic substrates at liquid helium temperatures. Results of her research have been published in several journals.

Dr. Godwin joined NASA in 1980, in the Payload Operations Division, Mission Operations Directorate, where she worked in payload integration (attached payloads and Spacelabs), and as a flight controller and payloads officer on several Shuttle missions.

Selected by NASA as an astronaut candidate in June 1985, Dr. Godwin became an astronaut in July 1986.

Her technical assignments have included working with flight software verification in the Shuttle Avionics Integration Laboratory (SAIL), and coordinating mission development activities for the Inertial Upper Stage (IUS), deployable payloads, and Spacelab missions. She also has served as Chief of Astronaut Appearances, Chief of the Mission Development Branch of the Astronaut Office and as the astronaut liaison to its Educational Working Group, Deputy Chief of the Astronaut Office, and Deputy Director, Flight Crew Operations Directorate.

Space Flight Experience:

A veteran of four space flights, Dr. Godwin has logged over 38 days in space, including over 10 EVA hours in two spacewalks. In 1991 she served as a Mission Specialist on STS-37, was the Payload Commander on STS-59 in 1994, flew on STS-76 in 1996, a *Mir* docking mission, and served on STS-108/International Space Station Flight UF-1 in 2001.

STS-37 *Atlantis* (April 5-11, 1991) was launched from the Kennedy Space Center, Florida, and returned to land at Edwards Air Force Base, California. During the 93 orbits of the mission, the crew deployed the Gamma Ray Observatory (GRO) to study gamma ray sources in the universe. GRO, at almost 35,000 pounds, was the heaviest payload deployed to that date by the Shuttle Remote Manipulator System (RMS). The crew also conducted an unscheduled spacewalk to free the GRO high gain antenna, and conducted the first scheduled extravehicular activity in 5-1/2 years to test concepts for moving around large space structures. Several mid-deck experiments and activities were conducted including test of elements of a heat pipe to study fluid transfer processed in microgravity environments (SHARE), a chemical processing apparatus to characterize the structure of biological materials (BIMDA), and an experiment to grow larger and more perfect protein crystals than can be grown on the ground (PCG II). *Atlantis* carried amateur radio equipment for voice contact, fast scan and slow scan TV, and packet radio. Several hundred contacts were made with amateur radio operators around the world. Mission duration was 143 hours, 32 minutes, 44 seconds.

STS-59 *Endeavour* (April 9-20, 1994) was the Space Radar Laboratory (SRL) mission. SRL consisted of three large radars, SIR-C/X-SAR (Shuttle Imaging Radar C/X-Band Synthetic Aperture Radar), and a carbon monoxide sensor that were used to enhance studies of the Earth's surface and atmosphere. The imaging radars operated in three frequencies and four polarizations. The multispectral capability of the radars provided information about the Earth's surface over a wide range of scales not discernible with previous single-frequency experiments. The carbon monoxide sensor MAPS (Measurement of Air Pollution by Satellite) used gas filter radiometry to measure the global distribution of CO in the troposphere. Real-time crew observations of surface phenomena and climatic conditions augmented with over 14,000 photographs aided investigators in interpretation and calibration of the data. The mission concluded with a landing at Edwards AFB after orbiting the Earth 183 times in 269 hours, 29 minutes.

STS-76 *Atlantis* (March 22-31, 1996) was the third docking mission to the Russian space station *Mir*. Following rendezvous and docking with *Mir*, transfer of a NASA astronaut to *Mir* for a 5-month stay was accomplished to begin a continuous presence of U.S. astronauts aboard *Mir* for the next two-year period. The crew also transferred 4800 pounds of science and mission hardware, food, water and air to *Mir* and returned over 1100 pounds of U.S. and ESA science and Russian hardware. Dr. Godwin performed a six-hour spacewalk, the first while docked to an orbiting space station, to mount experiment packages on the *Mir* docking module to detect and assess debris and contamination in a space station environment. A future shuttle mission was to retrieve the packages. The Spacehab module carried in the Shuttle payload bay was utilized extensively for transfer and return stowage of logistics and science and also carried Biorack, a small multipurpose laboratory used during this mission for research of plant and animal cellular function. This mission was also the first flight of Kidsat, an electronic camera controlled by classroom students via a Ku-band link between JSC Mission Control and the Shuttle, which uses digitized photography from the Shuttle for science and education. The STS-76 mission was accomplished in 145 orbits of the Earth, traveling 3.8 million miles in 221 hours and 15 minutes.

STS-108 *Endeavour* (December 5-17, 2001) was the 12th shuttle flight to visit the International Space Station. Endeavour's crew delivered the Expedition-4 crew and returned the Expedition-3 crew. The crew unloaded over 3 tons of supplies, logistics and science experiments from the Raffaello Multi-Purpose Logistics Module and repacked over 2 tons of items no longer needed on the station for return to Earth. Dr. Godwin used the Shuttle's robotic arm to install the MPLM onto the Station Node, and participated in a space walk to wrap thermal blankets around ISS Solar Array Beta Gimbal Assemblies. STS-108 was accomplished in 185 Earth orbits, traveling 4.8 million miles in 283 hours and 36 minutes.

Organizations:

Dr. Godwin is a member of the American Physical Society, the Ninety-Nines, Inc., Association of Space Explorers, Aircraft Owners and Pilots Association.

Special Honors:

Dr. Godwin is a recipient of NASA Outstanding Performance Rating, Sustained Superior Performance Award, and Outstanding Leadership Award.

Astronaut Biography of Helen Patricia Sharman

Personal Information:

Born 30 May 1963 in Sheffield, England.

When Helen was in school she enjoyed sports, French, German and science and also liked cycling and playing the piano.

One evening in June 1989, Helen heard a radio advertisement for a British consortium looking for astronaut volunteers. She was excited at the prospect of flying into space.

"Immediately, that small dream that I had had when I was younger that maybe, just maybe, space travel was possible," she said at a commencement in 1998, "Suddenly that was there in front of me. What would it be like to sit on top of a rocket and wait for that launch? What would it be like for me, as a chemist, to be able to grow crystals, crystals that you can't grow on earth, but probably best of all, what was it going to be like floating about feeling weightless, and for those reasons, very selfish reasons, when I got home that night I applied for the job of astronaut."

Helen was chosen as one of four finalists out of 13,000 applicants. She was one of two chosen to train in Russia.

Helen enjoys art, music, and plays the piano. She also plays squash and runs in the local park.

"Science has enabled me to have some terrific jobs. I have enjoyed my life so far, although it is nothing like I imagined it would be, and I am sure that the future will be different again. Science opens up new opportunities every day."

Education:

Sharman graduated with a degree in chemistry from the university of Sheffield.

Experience:

Sharman's first job was with GEC in London, where she worked for 3 years on monitors for ships, aircraft, and hospital life support systems. She used all kinds of science to solve problems, not just chemistry.

Her next job was at the Mars Confectionery in England. "The Company wanted to make a new and delicious ice cream and I was to be part of the team that scaled up production from a few bars in the laboratory to tons of ice cream every day in the factory."

A little later at the Mars plant, "I worked in the chocolate department at Mars, investigating the properties of chocolate and using different ingredients and machinery to make chocolate more quickly, more cheaply, and with the same flavor."

Astronaut Experience:

For the Soviet space mission *Project Juno,* Sharman went through cosmonaut training at the Yuri Gagarin Cosmonaut Training Center in Star City, Russia. Sharman didn't know any Russian before she left, but learned it during her training from a professor who knew no English. They used French as a common language to teach each other.

Space Flight Experience:

Sharman holds the distinction of being the first British woman in space. On 18 May 1991, Sharman and her crewmates Anatoli Artsebarski, and Sergei Krikalev lifted off in the *Soyuz* TM12 and docked with Mir. They were all part of *Mir* Expedition EO-09. Her flight was supposed to be paid for by a British consortium. However, the sponsoring British consortium wasn't quite able to come up with the money. The flight continued at Soviet expense with very limited UK experiments. Sharman's experiments included how weightlessness affects people, plants, and the growth of crystals.

Sharman and crew landed in the *Soyuz* TM11 on 26 May 26 1991. The flight lasted 7 days, 21 hours, and 13 minutes.

Organizations:

Sharman is a fellow of the Royal Society of Chemistry, the Royal Aeronautical Society, and the British Interplanetary Society. She is a fellow at Sheffield Hallam University and a senior fellow at the University of Sheffield.

Special Honors:

Sharman was awarded the Order of the British Empire in 1992.

Astronaut Biography of Millie Hughes-Fulford

Personal Information:

Born 21 December 1945, in Mineral Wells, Texas.

Millie wanted to go into space since she was, "five or six, maybe when I became interested, I loved space flight. I loved space fiction. I just thought I should be off the world. At that age you know you're watching a story, but you kind of think it's based in real life. I wanted to go into space."

Science fiction helped interest her, "Yes. I watched Buck Rogers on channel eight out of Dallas that we could get in Mineral Wells. And they had a whole series of science fiction in the morning for kids."

"Well I wanted to be an astronaut until I was about 16. And that's when I realized that all the real astronauts were men. And that they were all fighter pilots and women couldn't be fighter pilots. So then I decided that I would just be a government scientist," Millie recalls. "Well you know scientists always save people, and they were good and honorable."

Millie has a naturally scientific mind and even the games she liked, such as Clue, were puzzles.

Millie never let others keep her back. "If you live in Mineral Wells Texas, you knew girls could only be nurses or schoolteachers. My Dad did believe that I could do anything that I wanted to do, and encouraged me that way. And my mother would keep telling me that I should take shorthand so that I could get a job. That's all they knew. It was good advice at the time. Who would have thought that a girl could do it?"

Millie is married to George Fulford, a retired Captain with United Airlines. Her daughter is married and lives in Mill Valley, California.

Being an astronaut is hard work, but Millie knows it's hard on the family too. She says, "You know, the spouse goes through so much. It's so hard for them. The family needs an award. They need to be recognized. Because we get to go and have lots of fun, and the families still have to deal with the press and the worry, and they don't get to fly."

"Someone once asked me if I'd go again. And I said, 'you should ask my family that and they would tell you not only 'no' but 'hell no'."

Millie's recreational interests include scuba diving, swimming, gardening, photography, computer graphics and boating.

Her hope for the future, "I would like to see the planet join together and quit fighting each other and do some joint missions in space." Millie would like to see us finish the space station and then go to Mars. "The space station should be an integral part of Mars planning."

"Take math", advises Millie. "Take the hard sciences, not the easy sciences. Don't give up. Do your homework. Don't let people discourage you. Don't let people bully you. Keep your eye on the goal."

Education:

Millie Hughes-Fulford graduated from Mineral Wells High, in 1962. She entered college at age 16 and received a Bachelor of Science degree in Chemistry and Biology from Tarleton State University in 1968, and a Ph.D. from Texas Womans University in 1972.

Experience:

After college, Dr. Hughes-Fulford began her graduate work by studying plasma chemistry at Texas Woman's University as a National Science Foundation Graduate Fellow from 1968-1971 and an American Association of University Women fellow from 1971-1972.

Upon completing her doctorate degree at TWU in 1972, Dr. Hughes-Fulford joined the faculty of Southwestern Medical School, University of Texas, Dallas as a postdoctoral fellow with Marvin D. Siperstein. Her research focused on regulation of cholesterol metabolism.

Dr. Hughes-Fulford has contributed over 70 papers and abstracts in the areas of bone and cancer growth regulation. She was a Major in the US Army Reserve until 1995.

Upon her return to Earth, Dr. Hughes-Fulford has served as the Scientific Advisor to the Under Secretary of the Department of Veterans Affairs for 3 years. Today, as a Professor at the University of California Medical Center at San Francisco, Dr. Hughes-Fulford continues her research at the Department of Veterans Affairs Medical Center, San Francisco as Director of the Laboratory for Cell Growth and Differentiation and as a principal medical investigator.

She is currently the Deputy Associate Chief of Staff at the VA Medical Center, San Francisco. In addition, she is continuing her studies of the control of human colorectal cancer growth with the VA and the regulation of bone growth with NASA. As she puts it, "I study bones in space flight as well as T-cells. And on the VA side I study prostate cancer."

She is also the Principle Investigator on a series of SpaceHab/Biorack experiments examining the regulation of osteoblast (bone cell) growth regulation. The experiment, OSTEO flew on STS-76, in March of 1996. The experiment OSTEOGENE flew on STS-81 in January 1997. Her experiment OSTEOMARS flew on STS-84 in May 1997. The experiments are designed to examine the activation of bone cell growth and induction of gene expression in microgravity. These studies should help scientists understand the osteoporosis that occurs in astronauts during spaceflight.

Space Flight Experience:

NASA selected Dr. Hughes-Fulford as a payload specialist in January 1983. She flew in June 1991 aboard STS-40 Spacelab Life Sciences (SLS 1), the first Spacelab mission dedicated to biomedical studies. The SLS-1 mission flew over 3.2 million miles in 146 orbits and its crew completed over 18 experiments during a 9-day period bringing back more medical data than any previous NASA flight. Mission duration was 218 hours, 14 minutes and 20 seconds.

Organizations:

Dr. Hughes-Fulford is a member of the American Association for the Advancement of Science, American Society for Gravitational Science and Biology, American Society for Bone and Mineral Research, American Society for Cell Biology and the Association of Space Explorers.

Special Honors:

She was named the Federal Employee of the Year Award for the Western Region in 1985, International Zontian in 1992 and Marin County Woman of the Year in 1993. Dr. Hughes-Fulford has also received the NASA Space Flight Medal in 1991.

Astronaut Biography of Tamara E. "Tammy" Jernigan

Personal Information:

Born 7 May 1959, in Chattanooga, Tennessee.

Tammy believes that astronauts will venture to the planet Mars sometime soon. "I think there is a kid in school right now who will be the first person on Mars," Tammy told a group of school children. *"It is your generation that will be the next explorers in space." In an interview with Spacekids.com, Tammy advised young people to, "Stay in school and always try your best."*

Tammy is married to Peter J.K. "Jeff" Wisoff. She enjoys volleyball, racquetball, softball, and flying. As an undergraduate, she competed in intercollegiate athletics on Stanford's varsity volleyball team.

Education:

Jernigan graduated from Santa Fe High School, Santa Fe Springs, California, in 1977. She received a Bachelor of Science degree in physics (with honors), and a Master of Science degree in engineering science from Stanford University in 1981 and 1983, a Master of Science degree in astronomy from the University of California-Berkeley in 1985, and a doctorate in space physics and astronomy from Rice University in 1988.

Experience:

After graduating from Stanford University, Jernigan served as a research scientist in the Theoretical Studies Branch at NASA Ames Research Center from June 1981 until July 1985. Her research interests have included the study of bipolar outflows in regions of star formation, gamma ray bursters, and shock wave phenomena in the interstellar medium.

Astronaut Experience:

Selected as an astronaut candidate by NASA in June 1985, Dr. Jernigan became an astronaut in July 1986. Her assignments since then have included: software verification in the Shuttle Avionics Integration Laboratory (SAIL); operations coordination on secondary payloads; spacecraft communicator (CAPCOM) in Mission Control for STS-30, STS-28, STS-34, STS-33, and STS-32; lead astronaut for flight software development; Chief of the Astronaut Office Mission Development Branch; Deputy Chief of the Astronaut Office.

Prior to STS-96 she served as the Assistant for Station to the Chief of the Astronaut Office, directing crew involvement in the development and operation of the Station.

Currently, Dr. Jernigan serves as the Lead Astronaut for Space Station external maintenance. She also formulates and advocates Astronaut Office EVA input into the design, maintenance, and operation of research and systems modules built by NASA's Italian partners.

Space Flight Experience:

A veteran of five space flights, Dr. Jernigan has logged over 1,512 hours in space, including an EVA totaling 7 hours and 55 minutes. She was a mission specialist on STS-40 (June 5-14, 1991) and STS-52 (October 22-November 1, 1992), was the payload commander on STS-67 (March 2-18, 1995), and again served as a mission specialist on STS-80 (November 19 to December 7, 1996) and STS-96 (May 27 to June 6, 1999).

STS-40 Spacelab Life Sciences (SLS-1) was a dedicated space and life sciences mission aboard Space Shuttle *Columbia*. During the nine-day flight crewmembers performed experiments, which explored how humans, animals and cells respond to microgravity and readapt to Earth's gravity on return. Other payloads included experiments designed to investigate materials science, plant biology and cosmic radiation. Mission duration was 218 hours, 14 minutes, 20 seconds. Landing was at Edwards Air Force Base, California.

STS-52 was also launched aboard Space Shuttle *Columbia*. During the 10-day flight, the crew deployed the Italian Laser Geodynamic Satellite (LAGEOS), which will be used to measure movement of the Earth's crust, and operated the U.S. Microgravity Payload 1 (USMP-1). Also, the Space Vision System (SVS), developed by the Canadian Space Agency, was tested by the crew using a small target assembly, which was released from the remote manipulator system. The SVS will be used for Space Station construction. In addition, the crew performed numerous other experiments encompassing the areas of geophysics, materials science, biological research and applied research for Space Station. Mission duration was 236 hours, 56 minutes 13 seconds. Landing was at Kennedy Space Center, Florida.

STS-67 Astro-2 mission aboard the Space Shuttle Endeavour was the second flight of the Astro observatory, a unique complement of three telescopes. During this record-setting 16-day mission, the crew conducted observations around the clock to study the far ultraviolet spectra of faint astronomical objects and the polarization of ultraviolet light coming from hot stars and distant galaxies. Mission duration was 399 hours and 9 minutes. Landing was at Edwards Air Force Base in California.

On STS-80 the crew aboard Space Shuttle *Columbia* successfully deployed and retrieved the Wake Shield Facility (WSF) and the Orbiting Retrievable Far and Extreme Ultraviolet Spectrometer (ORFEUS) satellites. The free-flying WSF created a super vacuum in its wake and grew thin film wafers for use in semiconductors and other high-tech electrical components. The ORFEUS instruments, mounted on the reusable Shuttle Pallet Satellite, studied the origin and makeup of stars. Her two planned spacewalks were lost due to a jammed outer hatch on the airlock. Mission duration was a record breaking 423 hours, 53 minutes.

STS-96 *Discovery* (May 27 to June 6, 1999) was a 10-day mission during which the crew performed the first docking to the International Space Station, and delivered 4 tons of logistics and supplies in preparation for the arrival of the first crew to live on the station early next year. The mission was accomplished in 153 Earth orbits, traveling 4 million miles in 235 hours and 13 minutes, during which Dr. Jernigan performed an EVA of 7 hours and 55 minute duration.

Organizations:

Dr. Jernigan is a member of the American Astronomical Association, the American Physical Society, the United States Volleyball Association, and a Lifetime Member of the Girl Scouts.

Special Honors:

Dr. Jernigan has twice received the Distinguished Service Medal in 1997 and in 2000. She also received the Lowell Thomas Award, Explorer's Club (2000); Group Achievement Award - EVA Development Test Team (1997); Fédération Aéronautique Internationale Vladimir Komorov Diploma (1997, 1996); Outstanding Leadership Medal (1996); Outstanding Performance Award (1993); Exceptional Service Medal (1993); Laurels Award, Aviation Week (1991). She is a five-time recipient of the NASA Space Flight Medal (2000,1996, 1995, 1992, 1991).

Astronaut Biography of Roberta Lynn Bondar

Personal Information:

Born 4 December 1945 in Sault St. Marie, Ontario, Canada.

Canada's first woman in space wanted to become an astronaut since she was a small child. Roberta dreamed of seeing the Earth and other planets from space. She looked up into the clear night skies of Northern Ontario and imaged she was Flash Gordon on an asteroid in search of Ming the Merciless. When she was eight years old, she built plastic models, rockets, space stations and satellites.

Roberta has certification in scuba diving, parachuting, and holds a private pilot's license. She enjoys photography, biking, hot air ballooning, roller-blading, and flying.

Education:

Dr. Roberta Bondar attended elementary and secondary school in Sault St. Marie, Ontario. She has Bachelor of Science degrees in zoology and agriculture, from the University of Guelph, 1968. She has a Masters of Science degree in experimental pathology, from the University of Western Ontario, 1971. Her Ph.D. is in neurobiology, from the University of Toronto, 1974. And her M.D. was earned in 1977 from McMaster University. She was admitted as a Fellow of the Royal College of Physicians and Surgeons of Canada as a specialist in neurology in 1981.

Experience:

Dr. Bondar is a neurologist and researcher. After interning in internal medicine at Toronto General Hospital, she completed post-graduate medical training in neurology at the University of Western Ontario. She also did post-graduate medical work in Neuro-opthalmology at Tuft's New England Medical Center in Boston and the Playfair Neuroscience Unit of Toronto Western Hospital. She studied carotid Doppler ultrasound and transcranial Doppler at the Pacific Vascular Institute in Seattle. She was appointed Assistant Professor of Medicine (Neurology) at the McMaster University, 1982-84. In 1988 she became a staff scientist at the Sunnybrook Medical Center, Toronto. Since then she has also become a visiting research scholar at a number of medical and Neurology facilities and universities.

Dr. Bondar was one of the six original Canadian astronauts selected in December 1983. She began astronaut training in February 1984. She served as chairperson of the Canadian Life Sciences Subcommittee for the Space Station from 1985 to 1989, and as a member of the Ontario Premier's Council on Science and Technology from 1988 to 1989. In early 1990, she was designated a prime Payload Specialist for the first International Microgravity Laboratory Mission (IML-1).

Dr. Roberta Bondar is principal investigator in Transcranial Doppler on patients with Orthostatic Intolerance, University of New Mexico, Deaconess Hospital, Boston. She is also principal investigator in Transcranial Doppler on astronauts before and after space flight, for the Johnson Space Center, Edwards Air Force Base, and Kennedy Space Center.

She is currently a distinguished Professor at the Center for Advanced Technology Education (CATE), Ryerson Polytechnic University, Toronto, Ontario. She is also a distinguished Professor at the Faculty of Kinesiology, University of Western Ontario, London, Ontario. And she is a visiting research scholar for the Department of Neurology, University of New Mexico; and visiting research scientist for the Universities Space Research Association, Johnson Space Center, Houston, Texas.

Dr. Bondar is also author of "Touching the Earth". The book profits go to Friends of the Environment Foundation a non-profit organization of Canada Trust.

Space Flight Experience:

Dr. Roberta Bondar is Canada's first woman in space. She flew on the space shuttle *Discovery* during Mission STS-42, January 22-30, 1992. She was a payload specialist performing life science and material science experiments in the Spacelab and on the mid-deck.

Organizations:

Organizations that Dr. Bondar is a member of include:
Fellow, Royal College of Physicians and Surgeons of Canada
Member, American Academy of Neurology
Canadian Aeronautics and Space Institute
Canadian Society of Aerospace Medicine
College of Physicians and Surgeons of Ontario
Canadian Stroke Society, Aerospace Medical Association
Albuquerque Aerostat Ascension Association
American Society for Gravitational and Space Biology
Association for Space Explorers
Canadian Society of Aerospace Medicine
Canadian Medical Association
Ontario Medical Association
Canadian Association of Sports Medicine

Special Honors:

Special honors include:
Officer of the Order of Canada
The Order of Ontario
Canada 125 Medal
NASA's Space Medal
Hubertus Strughold Award, Space Medicine Branch, Aerospace Medicine Association
Award of Merit, University of Western Ontario
Medaille de L 'Excellence, L 'Association des Medicins de Langue Francaise du Canada
1995 Women's Intercultural Network International Women's Day Award;
1993 Alumnus of the Year, University of Guelph
Honorary Life Member, Canadian Federation of University Women

Astronaut Biography of N. Jan Davis

Personal Information:

Born 1 November 1953, at Cocoa Beach, Florida, but considers Huntsville Alabama, to be her hometown.

"I grew up here in Huntsville, which is where Von Braun and the rocket team was. I went to school with a lot of their kids. So in the early '60's when we were going to the moon, I moved here and was a part of all of the testing they were doing here in Huntsville. You know, they tested all of the engines, and the whole town would vibrate. It was a really big deal in Huntsville," recalls Jan. "When I say the whole town, I mean the windows would shake and you just knew we were doing something to help us go to the Moon. It was really exciting. With every mission, and every launch, they would make a big deal about it at school. I guess I became interested in space when everybody else in the country did, but it was just a very real thing for me here."

Jan enjoyed science and math. "I had one science teacher that really influenced me and made it very interesting and fun for me. I think that made a difference. With math I just enjoyed the challenge of it. I ended up taking all the courses I could and they didn't have any courses left for me when I was a senior in high school, so I started taking calculus at the university when I was in high school. Just because I really enjoyed it."

As a girl, Jan enjoyed Girl Scouts, and while she was an astronaut at NASA, she was an assistant Girl Scout Troop leader. "We worked on a lot of science badges, astronomy, things like that. I was really trying to make those things interesting for them and make them fun. That was really great."

It wasn't until women were first selected as astronauts that Jan thought it was possible for her to be one too. "When I was growing up, there weren't any women astronauts so I didn't even think it was possible until they started selecting women in 1978. So that's when I started thinking about it. And I went to work for NASA in 1979. So all of that kind of happened at the same time. So that's how I first got the idea."

And once she had wanted to be an astronaut, she kept at it. "Sometimes I would go to the IMAX movie 'The Dream is Alive' and watch that. I must have watched it ten times. Every time I got discouraged, I could go over there and that would motivate me and make me think, 'well it's tough now but I've got to keep going because one of these days maybe this dream will come true.'"

Jan enjoys flying, ice-skating, snow skiing, water sports, and needlepoint.

"My passion in the space industry is doing scientific research, and now that's on the space station. I think that's the real reason we're on the ISS," says Jan. "I hope we can really continue now that we have a permanent presence in space; we can do continuous research so that we can make some breakthroughs."

"I would encourage any woman to pursue her dream and to study whatever discipline she really enjoys," Jan told NASA's Women in Microgravity web site. "If she sets her goals, and perseveres, she can build the foundations to really make her dream come true. There are so many opportunities for women now, that gender should not be an issue. Be your very best at whatever you do!" This is good advice, whatever career you choose.

Education:
Jan Davis graduated from Huntsville High School in 1971. She received Bachelor of Science degrees in applied biology from Georgia Institute of Technology and in mechanical engineering from Auburn University in 1975 and 1977, respectively. She received a Master of Science degree and a doctorate in mechanical engineering from University of Alabama in Huntsville, in 1983 and 1985, respectively.

Experience:
After graduating from Auburn University in 1977, Dr. Davis joined Texaco in Bellaire, Texas, working as a petroleum engineer in tertiary oil recovery. She left in 1979 to work for NASA's Marshall Space Flight Center as an aerospace engineer.

In 1986, she was named as team leader in the Structural Analysis Division, and her team was responsible for the structural analysis and verification of the Hubble Space Telescope (HST), the HST maintenance mission, and the Advanced X-Ray Astrophysics Facility.

In 1987, she was assigned to be the lead engineer for the redesign of the solid rocket booster external tank attach ring for the space shuttle fleet. Dr. Davis did her graduate research at the University of Alabama in Huntsville, studying the long-term strength of pressure vessels due to the viscoelastic characteristics of filament-wound composites. She holds one patent, has authored several technical papers, and is a Registered Professional Engineer.

Astronaut Experience:
Dr. Davis became an astronaut in June 1987. Her initial technical assignment was in the Astronaut Office Mission Development Branch, where she provided technical support for Shuttle payloads. She served as a CAPCOM in Mission Control communicating with Shuttle crews on seven missions.

After her first space flight, Dr. Davis served as the Astronaut Office representative for the Remote Manipulator System (RMS), with responsibility for RMS operations, training, and payloads.

After her second space flight, she served as the Chairperson of the NASA Education Working Group and as Chief for the Payloads Branch, which provided Astronaut Office support for all Shuttle and ISS payloads.

After her flight on STS-85, Dr. Davis was assigned to NASA Headquarters as the Director of the Human Exploration and Development of Space (HEDS), Independent Assurance Office for the Office of Safety and Mission Assurance. In that position, Dr. Davis managed and directed independent assessments for the programs and projects assigned to the HEDS enterprise.

In July 1999, she was named the Deputy Director of the Flight Projects Directorate at the Marshall Space Flight Center. "My job primarily is involving the International Space Station, both with hardware and our payload operations center here. We have a control center that talks to the crew on the station about all the science and payloads that are on the space station," she explains. "We're actually building some hardware that goes on the space station, which is part of the job. But the operations center involves managing the science that's being done. So we work with the crew, timeline their activities, train them, and command experiments, on and off, from the ground. So we have that control center here, 24 hours a day."

Space Flight Experience:

A veteran of three space flights, Dr. Davis has logged over 673 hours in space. She flew as a mission specialist on STS-47 in 1992 and STS-60 in 1994, and was the payload commander on STS-85 in 1997. STS-47, Spacelab-J, was the 50th Space Shuttle mission. Launched on September 12, 1992, this cooperative venture between the United States and Japan, conducted 43 experiments in life sciences and materials processing. During the eight-day mission, she was responsible for operating Spacelab and its subsystems and performing a variety of experiments. After completing 126 orbits of the Earth, STS-47 Endeavour landed at Kennedy Space Center on September 20, 1992. STS-60 was the second flight of Spacehab (Space Habitation Module) and the first flight of the Wake Shield Facility (WSF). Launched on February 3, 1994, this flight was the first Shuttle flight on which a Russian Cosmonaut was a crewmember. During the eight-day mission, her prime responsibility was to maneuver the WSF on the RMS, to conduct thin film crystal growth and she was also responsible for performing scientific experiments in the Spacehab. The STS-60 *Discovery* landed at Kennedy Space Center on February 11, 1994, after completing 130 orbits of the Earth. Dr. Davis was the payload commander for STS-85, which was launched on *Discovery* on August 7, 1997. During this 12-day mission, Dr. Davis deployed and retrieved the CRISTA-SPAS payload, and operated the Japanese Manipulator Flight Demonstration (MFD) robotic arm. The mission also included several other scientific payloads for the conduct of research on astronomy, Earth sciences, life sciences, and materials science. The mission was accomplished in 189 Earth orbits, traveling 4.7 million miles. The STS-85 *Discovery* landed at Kennedy Space Center on August 19, 1997.

Organizations:

Dr. Davis is a Fellow of the American Society of Mechanical Engineers. She is also a Member of Tau Beta Pi, Omicron Delta Kappa, Pi Tau Sigma, and Sigma Gamma Tau honoraries, and Alpha Xi Delta social sorority.

Special Honors:

Dr. Davis has received the NASA Outstanding Leadership Medal (1998), NASA Exceptional Service Medal (1995), NASA Space Flight Medal (1992, 1994, 1997), Marshall Space Flight Center Director's Commendation (1987), NASA Fellowship for Full-Time Study (1983), ASME National Old Guard Prize (1978), and the Alpha Xi Delta Woman of Distinction (1993).

Astronaut Biography of Mae Jemison

Personal Information:

Born 17 October 1956 in Decatur, Alabama. Mae considers Chicago, Illinois, to be her hometown where she moved with her parents when she was four.

Mae was the first African-American woman astronaut.

"I'd wanted to be in the astronaut program since I was a child," Mae remembers in her book. "I remember when I was a child watching the *Mercury* and *Gemini* launches. I had a school teacher who would play them on television."

In her book, Mae recalls that in kindergarten she told her teacher that she wanted to be a scientist. The teacher replied, "Don't you mean a nurse?" and Mae corrected her saying, "No, I mean a scientist."

Mae says she dreamed of being in space for many years. "I really didn't have a concrete idea of what experiment you ran. But I saw myself working in that environment, and definitely had fantasies of being on another planet or on the Moon."

Mae loves to dance. When she was a girl, she would put on dance shows at family gatherings and do the latest popular dances, the "Four Corners", "Bugaloo", "Mashed Potato", "Popcorn", and "Skate".

Growing up in Woodlawn, part of Chicago's inner city, her brother had a close encounter with a gang member, and Mae witnessed her mother chase them off with a gun. After that incident, the whole family moved to the far south side of Chicago and became the first black family on their block.

Mae never lost her desire to go into space. "I was a senior in college when they opened up the Shuttle program to civilians. I said, 'yeah, I'll be able to apply because I'll have the qualifications in so many years' And that's when I made the very conscious decision to apply at some point."

Mae watched *Star Trek* as a child and it captured her imagination and respect. "Somewhere, someone else believed that other kinds of people would populate spaceships from Earth."

Her favorite *Star Trek* characters were Mr. Spock and Lt. Uhura. "I believe she was the first woman to appear regularly on television in a technical role. She was intelligent, skilled, gorgeous, cool, and looked a bit like me and the women around me. She had a wonderful soothing voice and manner. I was disappointed that she did not get more airtime."

In 1993, at the invitation of actor LeVar Burton, Mae appeared in the *Star Trek: The Next Generation* episode *Second Chances* as Lieutenant Palmer. She became the only *Enterprise* crewmember to have actually gone into space.

Mae Jemison once said, "Don't be limited by others' limited imaginations."

Mae's recreational interests include traveling, graphic arts, photography, sewing, skiing, collecting African Art, languages (Russian, Swahili, Japanese), and weight training. She has an extensive dance and exercise background and is an avid reader.

Education:

Mae Jemison graduated from Morgan Park High School, Chicago, Illinois, in 1973. She received a Bachelor of Science degree in chemical engineering, and fulfilled the requirements for a B.A. in African and Afro-American Studies, from Stanford University in 1977, and a doctorate degree in medicine from Cornell University in 1981.

Experience:

Dr. Jemison has a background in both engineering and medical research. She has worked in the areas of computer programming, printed wiring board materials, nuclear magnetic resonance spectroscopy, computer magnetic disc production, and reproductive biology.

Dr. Jemison completed her internship at Los Angeles County/USC Medical Center in July 1982 and worked as a General Practitioner with INA/Ross Loos Medical Group in Los Angeles until December 1982.

From January 1983 through June 1985, Dr. Jemison was the Area Peace Corps Medical Officer for Sierra Leone and Liberia in West Africa. Her task was managing the health care delivery system for U.S. Peace Corps and U.S. Embassy personnel including provision of medical care, supervision of the pharmacy and laboratory, medical administrative issues, and supervision of medical staff.

She developed a curriculum and taught volunteer personal health training, wrote manuals for self-care, developed and implemented guidelines for public health/safety issues for volunteer job placement and training sites. Dr. Jemison developed and participated in research projects on Hepatitis B vaccine, schistosomaisis and rabies in conjunction with the National Institutes of Health and the Center for Disease Control.

On return to the United States, Dr. Jemison joined CIGNA Health Plans of California in October 1985 and was working as a General Practitioner and attending graduate engineering classes in Los Angeles when she was selected to the astronaut program.

After being an astronaut, Dr. Jemison started The Earth We Share, an international science camp for kids ages 12 to 16. At camp, kids from different countries spend four weeks learning from each other and trying to answer global questions like how many people can the Earth hold. She says, "One of the things I thought of when I was in space is that the Earth will be here for years and years to come. We need to make sure it's still hospitable for us. We're not saving the Earth, we're saving ourselves."

Dr. Jemison is currently the director for the Jemison Institute for Advancing Technology in Developing Countries and a Professor of Environmental Studies at Dartmouth College. Her goal is to improve the quality of human life and ensure that future generations can grow and prosper.

Astronaut Experience:

Dr. Jemison was selected for the astronaut program in June 1987. Her technical assignments at NASA included: launch support activities at the Kennedy Space Center in Florida; verification of Shuttle computer software in the Shuttle Avionics Integration Laboratory (SAIL); and Science Support Group activities.

Dr. Jemison was the science mission specialist on STS-47 Spacelab-J (September 12-20, 1992). STS-47 was a cooperative mission between the United States and Japan. The eight-day mission was accomplished in 127 orbits of the Earth, and included 44 Japanese and U.S. life science and materials processing experiments. Dr. Jemison was a co-investigator on the bone cell research experiment flown on the mission. The Endeavour and her crew launched from and returned to the Kennedy Space Center in Florida. In completing her first space flight, Dr. Jemison logged 190 hours, 30 minutes, 23 seconds in space.

Dr. Jemison left NASA in March 1993.

Organizations:

Dr. Jemison is a member of the American Chemical Society, Association for the Advancement of Science, and the Association of Space Explorers. She is a board member for the World Sickle Cell Foundation, and the American Express Geography Competition. She is an Honorary Board Member for the Center for the Prevention of Childhood Malnutrition and a Clinical Teaching Associate for the University of Texas Medical Center.

Special Honors:

Some of Mae Jemison's honors include:

National Achievement Scholarship (1973-1977)
Stanford representative to Carifesta '76 in Jamaica
1979 CIBA Award for Student Involvement
American Medical Student Association (AMSA) study group to Cuba
Grant from International Travelers Institute for health studies in rural Kenya (1979)
Recipient of Essence Award (1988), and Gamma Sigma Gamma Woman of the Year (1989)

Astronaut Biography of Susan J. Helms

Personal Information:

Born 26 February 1958, in Charlotte, North Carolina, but considers Portland, Oregon, to be her hometown.

Susan is the first American military woman astronaut.

Susan always wanted to fly in the Air Force. "Definitely my interest in the Air Force has been there since probably the day I was born [because] my dad was career Air Force," Susan told a NASA reporter. "And I basically decided at a young age - and by young, I mean junior high - that the whole thing of being in the military and particularly the Air Force appealed to me. There were a couple of strong points that I wanted. One of them was the opportunity to travel - little did I know how far that would go - and then also the opportunity for a stable career. I like the idea of moving every few years and seeing different places, and it appeared that my dad had had a very rewarding Air Force career. It looked like, as an engineer, I could also have the same."

Susan became as astronaut to fly higher and faster. "I basically had been working on some jobs in the Air Force that allowed me to fly in jets. When I had a taste of that, I got fairly addicted to it, and at some point along that career path, I decided, 'Well, how can I fly higher and faster than I am doing right now?' And that, combined with a couple of other brushes with NASA people, made me realize that this could be something I'd be very interested in doing, and I just put my application in, of course, never expecting to get selected. But when that happened, my Air Force career sort of took a turn to a NASA career, and I've had the chance to do just that: fly higher and faster. And that was what I was originally after."

"My dad and my mom are both very influential. I would say the other people that influenced me were teachers," says Susan. "When I was in junior high, I had some very influential math and science teachers - interestingly, for the most part they were all female - a very strong guidance counselor, and a lot of very supportive friends."

Susan enjoys piano and other musical activities, jogging, traveling, reading, computers, and cooking.

Susan plays keyboard for MAX-Q, a rock-n-roll band. As she once told *SPACE.com* about the band, "If it's easy, we play it." Musicians in the astronaut band include Carl Walz, lead vocalist; Hoot Gibson, lead guitar; Kevin Chilton, guitar; Jim Wetherbee, drums; Susan Helms, keyboard; and Chris Hadfield, bass.

Education:

Susan Helms graduated from Parkrose Senior High School, Portland, Oregon, in 1976. She received a Bachelor of Science degree in aeronautical engineering from the U.S. Air Force Academy in 1980, and a Master of Science degree in aeronautics/astronautics from Stanford University in 1985.

Experience:

Colonel Helms graduated from the U.S. Air Force Academy in 1980. She received her commission and was assigned to Eglin Air Force Base, Florida, as an F-16 weapons separation engineer with the Air Force Armament Laboratory.

In 1982, she became the lead engineer for F-15 weapons separation. In 1984, she was selected to attend graduate school. She received her degree from Stanford University in 1985 and was assigned as an assistant professor of aeronautics at the U.S. Air Force Academy.

In 1987, Colonel Helms attended the Air Force Test Pilot School at Edwards Air Force Base, California. After completing one year of training as a flight test engineer, Helms was assigned as a USAF Exchange Officer to the Aerospace Engineering Test Establishment, Canadian Forces Base, Cold Lake, Alberta, Canada, where she worked as a flight test engineer and project officer on the CF-18 aircraft.

She was managing the development of a CF-18 Flight Control System Simulation for the Canadian Forces when selected for the astronaut program.

As a flight test engineer, Helms has flown in 30 different types of U.S. and Canadian military aircraft.

Space Flight Experience:

Selected by NASA in January 1990, Helms became an astronaut in July 1991. She flew on STS-54 (1993), STS-64 (1994), STS-78 (1996), STS-101 (2000) and served aboard the International Space Station as a member of the Expedition-2 crew (2001). A veteran of five space flights, Helms has logged over 206 days in space, including an EVA of 8 hours and 56 minutes (world record).

STS-54 *Endeavour*, January 13-19, 1993. The primary objective of this mission was the deployment of a $200-million NASA Tracking and Data Relay Satellite (TDRS-F). A Diffuse X-Ray Spectrometer (DXS) carried in the payload bay, collected over 80,000 seconds of quality X-ray data that will enable investigators to answer questions about the origin of the Milky Way galaxy. The crew demonstrated the physics principles of everyday toys to an interactive audience of elementary school students across the United States. A highly successful Extravehicular Activity (EVA) resulted in many lessons learned that benefit the Space Station *Freedom* assembly. Mission duration was 5 days, 23 hours, 38 minutes, 17 seconds.

STS-64 *Discovery*, September 9-20, 1994. On this flight, Helms served as the flight engineer for orbiter operations and the primary RMS operator aboard space shuttle. The major objective of this flight was to validate the design and operating characteristics of Lidar in Space Technology Experiment (LITE) by gathering data about the Earth's troposphere and stratosphere. Additional objectives included the deployment and retrieval of SPARTAN-201, a free-flying satellite that investigated the physics of the solar corona, and the testing of a new EVA maneuvering device. The Shuttle Plume Impingement Flight Experiment (SPIFEX) was used to collect extensive data on the effects of jet thruster impingement, in preparation for proximity tasks such as space station docking. Mission duration was 10 days, 22 hours, 51 minutes.

STS-78 *Columbia*, June 20 to July 7, 1996, Helms was the payload commander and flight engineer aboard *Columbia*, on the longest Space Shuttle mission to date. The mission included studies sponsored by ten nations and five space agencies, and was the first mission to combine both a full microgravity studies agenda and a comprehensive life science investigation. The Life and Microgravity Spacelab mission served as a model for future studies on board the International Space Station. Mission duration was 16 days, 21 hours, 48 minutes.

STS-101 *Atlantis*, May 19-29, 2000, was a mission dedicated to the delivery and repair of critical hardware for the International Space Station. Helms prime responsibilities during this mission were to perform critical repairs to extend the life of the Functional Cargo Block (FGB). In addition, she had prime responsibility of the onboard computer network and served as the mission specialist for rendezvous with the ISS. Mission duration was 9 days, 20 hours and 9 minutes.

Helms lived and worked onboard the International Space Station as a member of the second crew to inhabit the International Space Station *Alpha*. The Expedition-2 crew (two American astronauts and one Russian cosmonaut) launched on March 8, 2001 onboard STS-102 *Discovery* and successfully docked with the station on March 9, 2001. The Expedition-2 crew installed and conducted tests on the Canadian made Space Station Robotic arm (SSRMS), conducted internal and external maintenance tasks (Russian and American), in addition to medical and science experiments. During her stay onboard the Space Station, Helms installed the Airlock (brought up on the STS-104 mission) using the SSRM. She and her crewmates also performed a 'fly around' of the Russian *Soyuz* spacecraft and welcomed the visiting *Soyuz* crew that included the 1st Space Tourist Dennis Tito. On March 11 she performed a world record 8 hour and 56 minute space walk to install hardware to the external body of the laboratory module. Helms spent a total of 163 days aboard the Space Station. She returned to earth with the STS-105 crew aboard *Discovery* on August 22, 2001.

Organizations:

Helms is a member of Women Military Aviators, U.S. Air Force Academy Association of Graduates, Stanford Alumni Association, Association of Space Explorers, Sea/Space Symposium, and a Chi Omega Alumni.

Special Honors:

Colonel Helms is a recipient of the Distinguished Superior Service Medal, the Defense Meritorious Service Medal, the Air Force Meritorious Service Medal, the Air Force Commendation Medal, NASA Space Flight Medals, and the NASA Outstanding Leadership Medal. She was also named a Distinguished Graduate of the USAF Test Pilot School, and recipient of the R.L. Jones Award for Outstanding Flight Test Engineer, Class 88A. In 1990, she received the Aerospace Engineering Test Establishment Commanding Officer's Commendation, a special award unique to the Canadian Forces. Named the Air Force Armament Laboratory Junior Engineer of the Year in 1983.

Astronaut Biography of Ellen Ochoa

Personal Information:

Born 10 May 1958 in Los Angeles, California, but considers La Mesa, California, to be her hometown.

Ellen told *Scholastic.com* that her Hispanic roots come from her father's side. "His parents were Mexican, but my father was born in this country. He was one of 12 children. My father grew up speaking both Spanish and English, but unfortunately he didn't speak Spanish with us at home. When I was growing up, my father believed, as many people did at the time, that there was a prejudice against people speaking their native language. It's really too bad, and I'm glad that things have changed in recent years."

Ellen is one of five children. While in junior high school, her parents divorced and she moved with her mother to La Mesa California.

Ellen is married to Coe Fulmer Miles of Molalla, Oregon. They have two sons.

"I don't think you have to be a special person to go into math, science or technical careers," says Ellen. "I didn't initially consider the sciences or becoming an astronaut because I didn't know any engineers or scientists."

That changed in college, and after Ellen graduated with honors she realized that she could become an astronaut. "I was in graduate school when I first thought about being an astronaut. I applied because some friends who were applying made me aware of the opportunity with NASA."

What does she think it takes to be an astronaut? "Good organizational skills are especially important because astronauts must learn a lot of material quickly, both in the early, training phase and throughout their careers—from workbooks to lectures, to practice in simulators."

Ellen wasn't selected as an astronaut on her first attempt. In 1985 she applied and interviewed but wasn't chosen. The fact that she was selected later is a lesson to keep striving for what you want. "I would advise everyone to set their aspirations high and then shoot for them. I don't think it matters if you reach that one lofty goal. In reaching high, you will encounter other opportunities that can lead to interesting career paths and an exciting life."

Ellen is a classical flautist and private pilot, and also enjoys volleyball and bicycling. She took her flute into space on her first flight.

Education:

Ellen Ochoa graduated from Grossmont High School, La Mesa, California, in 1975. She received a Bachelor of Science degree in physics from San Diego State University in 1980, and a Master of Science degree and doctorate in electrical engineering from Stanford University in 1981 and 1985, respectively.

Experience:

As a doctoral student at Stanford, and later as a researcher at Sandia National Laboratories and NASA Ames Research Center, Dr. Ochoa investigated optical systems for performing information processing. She is a co-inventor on three patents for an optical inspection system, an optical object recognition method, and a method for noise removal in images. As Chief of the Intelligent Systems Technology Branch at Ames, she supervised 35 engineers and scientists in the research and development of computational systems for aerospace missions. Dr. Ochoa has presented numerous papers at technical conferences and in scientific journals.

Selected by NASA in January 1990, Dr. Ochoa became an astronaut in July 1991. Her technical assignments to date include flight software verification, crew representative for flight software and computer hardware development, crew representative for robotics development, testing, and training, Assistant for Space Station to the Chief of the Astronaut Office, directing crew involvement in the development and operation of the Station, and spacecraft communicator (CAPCOM) in Mission Control. A veteran of three space flights, Dr. Ochoa has logged over 719 hours in space.

Space Flight Experience:

Ellen Ochoa is the first Hispanic woman in space. She was a mission specialist on STS-56 in 1993, was the Payload Commander on STS-66 in 1994, and was a mission specialist and flight engineer on STS-96 in 1999. Dr. Ochoa was a flight engineer on STS-110.

STS-56 ATLAS-2 *Discovery* (April 4-17, 1993) was a 9-day mission during which the crew conducted atmospheric and solar studies in order to better understand the effect of solar activity on the Earth's climate and environment. Dr. Ochoa used the Remote Manipulator System (RMS) to deploy and capture the Spartan satellite, which studied the solar corona.

Dr. Ochoa was the Payload Commander on the STS-66 Atmospheric Laboratory for Applications and Science-3 mission (November 3-14, 1994). ATLAS-3 continued the series of Spacelab flights to study the energy of the Sun during an 11-year solar cycle and to learn how changes in the sun's irradiance affect the Earth's climate and environment. Dr. Ochoa used the RMS to retrieve the CRISTA-SPAS atmospheric research satellite at the end of its 8-day free flight.

STS-96 *Discovery* (May 27 to June 6, 1999) was a 10-day mission during which the crew performed the first docking with the International Space Station, and delivered 4 tons of logistics and supplies in preparation for the arrival of the first crew to live on the station early the next year. Dr. Ochoa coordinated the transfer of supplies and also operated the RMS during the 8-hour space walk. The mission was accomplished in 153 Earth orbits, traveling 4 million miles in 235 hours and 13 minutes.

STS-110, *Atlantis* (April 8-19, 2002) was a 10-day mission to the International Space Station. The mission installed and outfitted the S-zero truss and prepared the mobile transporter for use. Mission Specialist Ellen Ochoa and Expedition Four Flight Engineer Dan Bursch operated the station's robot arm.

Organizations:

Ochoa is a member of the Optical Society of America (OSA), the American Institute of Aeronautics and Astronautics (AIAA), Phi Beta Kappa and Sigma Xi honor societies. She is also a Member of the Presidential Commission on the Celebration of Women in American History.

Special Honors:

Her NASA awards include the Exceptional Service Medal (1997), Outstanding Leadership Medal (1995), Space Flight Medals (1999, 1994, 1993), and two Space Act Tech Brief Awards (1992). She is also the recipient of numerous other awards, including the Women in Aerospace Outstanding Achievement Award, The Hispanic Engineer Albert Baez Award for Outstanding Technical Contribution to Humanity, the Hispanic Heritage Leadership Award, and San Diego State University Alumna of the Year.

Astronaut Biography of Nancy Jane Currie

Personal Information:

Born 29 December 1958, in Wilmington, Delaware, but considers Troy, Ohio, to be her hometown.

"I'd say from a very early age, I knew I wanted to fly," Nancy told ArmyLINK News. "I mean, I just dreamed about flying probably from the time I could walk. But she didn't expect to be an army pilot and then an astronaut. "I think it's kind of interesting, especially for a woman my age, because when we were kids growing up, women weren't military pilots."

Her primary job on Shuttle missions is to control the shuttle's robotic arm. How she controls the shuttle's arm is a lot like flying. "If you ask me when I'm hovering a helicopter, 'What inputs are you putting in?' I couldn't tell you because I'm just responding to the aircraft. Same thing here. I'm just responding to where I know they need to be placed and flying them there."

Nancy's robotic arm flying technique is a little different. "I tend not to try to fly the arm on the fastest mode possible. I actually don't think that's a very good way to fly the arm, although some people do. My technique instead is to be very deliberate in my movements but to know exactly where they're going so that when I stop and say, 'How does it look?' they say, 'It looks great.' They start working. They don't say, 'Move me three inches up,' 'Move me a foot over here,' whatever."

Nancy is married to David W. Currie. They have one daughter.

She enjoys weight lifting, running, swimming, scuba diving, and skiing.

Education:

Nancy Currie graduated from Troy High School, Troy, Ohio, in 1977. She received a Bachelor of Arts degree, with honors, in biological science from The Ohio State University, Columbus, Ohio, in 1980 and a Master of Science degree in safety from the University of Southern California in 1985. Her doctorate in industrial engineering was earned in 1997 from the University of Houston.

Experience:

Lieutenant Colonel Currie has served in the United States Army for over twenty years. Prior to her assignment at NASA in 1987, she attended initial rotary wing pilot training and was subsequently assigned as an instructor pilot at the U.S. Army Aviation Center.

In the Army, Lieutenant Colonel Currie has served in a variety of leadership positions including section leader, platoon leader, and brigade flight standardization officer. As a Master Army aviator she has logged over 3,900 flying hours in a variety of rotary-wing and fixed-wing aircraft.

Lieutenant Colonel Currie was assigned to NASA Johnson Space Center in September 1987 as a flight simulation engineer on the Shuttle Training Aircraft, a complex airborne simulator which models flight characteristics of the Orbiter.

An astronaut since 1990, she has been involved in robotic hardware and procedure development for the shuttle and space station and has worked as a spacecraft communicator.

Lieutenant Colonel Currie has also served as the chief of both the Astronaut Office Robotics and Payloads-Habitability branches.

Space Flight Experience:

A veteran of four space shuttle missions, Dr. Currie has accrued 1000 hours in space. She flew as mission specialist 2, flight engineer, on STS-57 (1993), STS-70 (1995), STS-88 (1998; the first International Space Station assembly mission), and STS-109 (2002).

STS-57 Endeavour (June 21 to July 1, 1993). The primary objective of this mission was the retrieval of the European Retrievable Carrier satellite (EURECA). Additionally, this mission featured the first flight of Spacehab, a commercially provided mid deck augmentation module for the conduct of microgravity experiments, as well as a spacewalk by two crew members, during which Dr. Currie operated the Shuttle's robotic arm. Spacehab carried 22 individual flight experiments in materials and life sciences research. STS-57 orbited the Earth 155 times and covered over 4.1 million miles in over 239 hours and 45 minutes.

STS-70 *Discovery* (July 13-22, 1995). The five-member crew deployed the final NASA Tracking and Data Relay Satellite to complete the constellation of NASA's orbiting communication satellite system. Dr. Currie also conducted a myriad of biomedical and remote sensing experiments. STS-70 orbited the Earth 143 times and covered over 3.7 million miles in over 214 hours and 20 minutes.

STS-88 Endeavour (December 4-15, 1998). STS-88, ISS Flight 2A was the first International Space Station assembly mission. The primary objective of this 12-day mission was to mate the first American-made module, Unity, to the first Russian-made module, Zarya. Dr. Currie's primary role was to operate the Shuttle's 50-foot robotic arm to retrieve Zarya and connect the first two station segments. Two crewmembers performed a series of three space walks to connect electrical umbilicals and to attach hardware to the exterior structure for use during future EVA's. Dr. Currie also operated the robot arm during the space walks. During the mission the STS-88 crew ingressed the International Space Station to complete systems activation and installation of communication's equipment. The crew also deployed two small satellites. STS-88 completed 185 orbits of the Earth and covered over 4.7 million miles in 283 hours and 18 minutes.

Dr. Currie loves being an astronaut. After STS-88, she told reporters, "Being on the first-ever station assembly mission and putting the first two pieces together and being one of the first individuals inside the new International Space Station — I mean those memories are just fantastic and I think some of us did think, well, 'What are we going to do to top this?'"

STS-109 *Columbia* (March 1-12, 2002). STS-109 was the fourth mission to service the Hubble Space Telescope. During the flight, Dr. Currie's primary role was to operate the Shuttle's 50-foot robot arm to retrieve and re-deploy the telescope following the completion of numerous upgrades and repairs. Dr. Currie also operated the robot arm during a series of five consecutive spacewalks performed by four crewmembers. Hubble's scientific capabilities and power system were significantly upgraded with the replacement of both solar arrays and the primary power control unit, the installation of the Advanced Camera for Surveys, and a scientific instrument cooling system. The Hubble Space Telescope was then boosted to a higher orbit and redeployed to continue its mission of providing views of the universe which are unmatched by ground-based telescopes or other satellites. STS-109 completed 165 earth orbits and covered over 3.9 million miles in over 262 hours.

Organizations:

Dr Currie is a member of Army Aviation Association of America, Phi Kappa Phi, Ohio State University and ROTC Alumni Associations, Institute of Industrial Engineers, and Human Factors and Ergonomics Society.

Special Honors:

Lieutenant Colonel Currie is a Distinguished Graduate of the Army Air Defense Artillery Officer Basic Course (1981); Honor Graduate of the Army Rotary Wing Aviator Course (1982) and the Army Aviation Officer Advanced Course

(1986); NASA Flight Simulation Engineering Award (1988); NASA Space Flight Medal (1993, 1995, 1998, 2002); Defense Superior Service Medal (1993, 1999); Ohio Veteran's Hall of Fame (1994); Troy, Ohio Hall of Fame (1996); Ohio State University Army ROTC Hall of Fame (1996); Silver Order of St. Michael, Army Aviation Award (1997).

Astronaut Biography of Janice Voss

Personal Information:

Born 8 October 1956, in South Bend, Indiana, but considers Rockford, Illinois, to be her hometown.

Janice first became interested in space in sixth grade. That's when she decided she wanted to be an astronaut.

She enjoyed reading science fiction books as a girl. It's been her favorite hobby ever since she read "A Wrinkle in Time" by Madeleine L'Engle's. After that she read books by Isaac Asimov, Robert Heinlein, and Arthur C. Clarke.

Janice enjoyed dancing and volleyball as a child.

Janice received encouragement from her parents, teachers, and bosses to become an astronaut. *When she was discouraged she sought advice from friends. To achieve her goal she says she, "Read about what other astronauts had done, and sought advice from astronauts."*

As an astronaut Janice enjoys, "being part of a really tight team, including flight controllers, trainers, and payload investigators." However, she continues, "Irregular and unpredictable hours make personal life more challenging." While in space, Janice, like other astronauts, really enjoys the view, but also likes to watch physics in action. Janice says the biggest drawback to living in space is personal hygiene like bathing and washing your hair. And of course being isolated from your friends and family is also difficult. She still enjoys reading, science fiction, dancing, volleyball, and flying.

Education:

Janice Voss graduated from Minnechaug Regional High School, Wilbraham, Massachusetts, in 1972. She received a Bachelor of Science degree in engineering science from Purdue University in 1975, a Master of Science degree in electrical engineering and a doctorate in aeronautics/astronautics from the Massachusetts Institute of Technology in 1977 and 1987, respectively. From 1973 to 1975 she took correspondence courses at the University of Oklahoma. She also did some of her graduate work in space physics at Rice University in 1977 and 1978.

Experience:

Dr. Voss was a co-op at the NASA Johnson Space Center from 1973 to 1975. During that time she did computer simulations in the Engineering and Development Directorate.

In 1977 she returned to the Johnson Space Center and, for a year, worked as a crew trainer, teaching entry guidance and navigation. She completed her doctorate in 1987 and accepted a job with Orbital Sciences Corporation. Her responsibilities there included mission integration and flight operations support for an upper stage called the Transfer Orbit Stage (TOS). TOS launched the Advanced Communications Technology Satellite (ACTS) from the Space Shuttle in September 1993, and the Mars Observer from a Titan in the fall of 1992.

Astronaut Experience:

Selected by NASA in January 1990, Dr. Voss became an astronaut in July 1991. She is qualified for flight assignment as a mission specialist. Her technical assignments have included working Spacelab/Spacehab issues for the Astronaut Office Mission Development Branch, and robotics issues for the Robotics Branch.

Space Flight Experience:

She served aboard STS-57 in 1993, STS-63 in 1995, STS-83 & STS-94 in 1997, and STS-99 in 2000. A veteran of five space flights, Dr. Voss has logged over 49 days in space, traveling 18.8 million miles in 779 Earth orbits.

Dr. Voss first flew on STS-57 (June 21 to July 1, 1993). Mission highlights included retrieval of the European Retrievable Carrier (EURECA) with the Shuttle's robotic arm, a spacewalk by two crewmembers, and an assortment of experiments in the first flight of the Spacehab mid-deck augmentation module.

She next flew on STS-63 (February 3-11, 1995). Mission highlights included the rendezvous with the Russian Space Station, Mir, the deployment and retrieval of Spartan 204, and the third flight of Spacehab.

She also flew as payload commander on STS-83 (Apr 4-8, 1997). The STS-83 Microgravity Science Laboratory (MSL-1) Spacelab mission was cut short because of problems with one of the Shuttle's three fuel cell power generation units.

The entire crew and payload re-flew on STS-94 (July 1-17, 1997). The STS-94 MSL-1 Spacelab mission focused on materials and combustion science research in microgravity.

Most recently, Voss served on STS-99 (February 11-22, 2000). This was an 11-day flight during which the international crew aboard Space Shuttle Endeavour worked dual shifts to support radar-mapping operations. The Shuttle Radar Topography Mission mapped more than 47 million miles of the Earth's land surface.

Organizations:

Dr. Voss is a member of the American Institute of Aeronautics and Astronautics (AIAA).

Special Honors:

Dr. Voss has received two NASA Space Flight Medals (1993, 1995), Zonta Amelia Earhart Fellowship (1982), Howard Hughes Fellowship (1981), and the National Science Foundation Fellowship (1976).

Astronaut Biography of Chiaki Mukai

Personal Information:

Born 6 May 1952, in Tatebayashi, Gunma Prefecture, Japan.

Chiaki was the first Japanese woman astronaut to fly in space.

John Glenn, a crewmember on her second flight, referred to Chiaki as the person most likely to win a sumo tournament in outer space. Not because of her figure though, he explained to Tokyo Classified, "She has more energy than anyone I know of."

On her second flight Chiaki talked with the Japanese prime minister. During the conversation, she read the beginning of a tanka poem that she had written describing her feeling of being in space *'Chuugaeri, Nandomo dekiru, Mujuuryoku'* ("Turn space somersaults, As many as you like, That is weightlessness").

There was a competition to complete the

poem, and Chiaki received over 144 thousand entries. Several were given awards, but the ending lines that she composed were *Chakuchi dekinai / kono modokashisa* ("But how unsettling not to be able to land on my feet").

"On Earth, we always fall to the ground," she explained. "But this wasn't the case in space, where it was so easy to do somersaults. Since very few people have actually experienced this, I might be the only one who would have finished the poem this way."

Chiaki's recreational interests include snow skiing, competitive Alpine skiing, bass fishing, scuba diving, tennis, golf, photography, American Literature, and traveling.

She is married to Makio Mukai.

"The fact that I am a woman has never occurred to me as either a limitation or advantage," Chiaki told a NASA web site. "If you have something you really want to accomplish, and believe you can do it well – based on your education and motivation – surpass any obstacles and challenges and go for it! If you can dream it, you can do it."

Education:

Chiaki Mukai graduated from Keio Girls' High School in Tokyo, in 1971. She received her doctorate in Medicine, Keio University School of Medicine in 1977, a doctorate in physiology, Keio University School of Medicine in 1988, and was board certified as a cardiovascular surgeon, Japan Surgical Society in 1989.

Experience:

From 1977 through 1978, Dr. Mukai worked as a resident in General Surgery, Keio University Hospital in Tokyo.

In 1989 she joined the Medical Staff in General Surgery, Shimizu General Hospital, Shizuoka Prefecture, and on the Medical Staff in Emergency Surgery, Saiseikai Kanagawa Hospital, Kanawaga Prefecture in 1979.

Dr. Mukai began her work as a resident in Cardiovascular Surgery, Keio University Hospital in 1980 and served on the Medical Staff in Cardiovascular Surgery, Saiseikai Utsunomiya Hospital, Tochigi Prefecture in 1982.

She returned to Keio University Hospital in 1983 as the Chief Resident in Cardiovascular Surgery, and was later promoted to Assistant Professor of the Department of Cardiovascular Surgery, Keio University.

Since 1992, Dr. Mukai has been a Research Instructor of the Department of Surgery, Baylor College of Medicine, Houston, Texas and a Visiting Associate Professor of the Department of Surgery, Keio University School of Medicine, Tokyo.

Astronaut Experience:

In 1985, Dr. Mukai was selected by the National Space Development Agency of Japan (NASDA), as one of three Japanese Payload Specialist candidates for the First Material Processing Test (Spacelab-J) which flew aboard STS-47. She also served as a back-up payload specialist for the Neurolab (STS-90) mission.

As a NASDA science astronaut, she became a visiting scientist of the Division of Cardiovascular Physiology, Space Biomedical Research Institute, and NASA Johnson Space Center from 1987 through 1988.

Space Flight Experience:

Dr. Mukai has logged over 566 hours in space. She flew aboard STS-65 in 1994 and STS-95 in 1998, and is the first Japanese female astronaut to fly in space. She is also the first female Japanese astronaut to fly in space *twice*.

STS-65 *Columbia* (July 8-23, 1994) was the second International Microgravity Laboratory (IML-2) flight. The mission consisted of 82 investigations of Space Life Science (Human Physiology, Space Biology, Radiation Biology, and Bioprocessing) and Microgravity Science (Material Science, Fluid Science and Research on the Microgravity Environment and Countermeasures). IML-2 was also designated as an extended duration orbit mission focusing on medical experiments related to the cardiovascular system, autonomic nerve system, and bone and muscle metabolism. The mission was accomplished in 236 orbits of the Earth, traveling over 6.1 million miles in 353 hours and 55 minutes.

STS-95 *Discovery* (October 29 to November 7, 1998) was a 9-day mission during which the crew supported a variety of research payloads including deployment of the Spartan solar-observing spacecraft, the Hubble Space Telescope Orbital Systems Test Platform, and investigations on space flight and the aging process. STS-95 is scheduled for launch in October 1998. The mission was accomplished in 134 Earth orbits, traveling 3.6 million miles in 213 hours and 44 minutes.

Organizations:

Dr. Mukai is a member of the American Aerospace Medical Association, Japan Society of Microgravity Applications, Japan Society of Aerospace and Environmental Medicine, Japanese Society for Cardiovascular and Thoracic Surgery, and the Japan Surgical Society.

Special Honors:

Dr. Mukai has received the Outstanding Service Award - The Society of Japanese Women Scientists (1996), Special Congressional Recognition - U.S. Congress (1995), Happy Hands Award - Satte Junior Chamber of Commerce (1995), Aeromedical Association of Korea Honorary Membership (1995), Tatebayashi Children's Science Exploratorium Honorary President (1995), Prime Minister's Special Citation for Contributions to Gender Equality (1995), The De La Vaux Medal - The Federation Aeronautique Internationale (1995), The Award for Distinguished Service in Advancement of Space Biology - Japanese Society for Biological Sciences in Space (1995), Prime Minister's Special Citation (1994), Minister of State for Science and Technology's Commendation (1994 & 1992), People of Gunma Prefecture's Certificate of Appreciation (1994), Honorary Citizen of Tatebayashi City (1994), Outstanding Service Award - National Space Development Agency of Japan (1994 & 1992), Award for Distinguished Accomplishments - Tokyo Women's Foundation (1994) and a Commendation for Technology - Japan Society of Aeronautical and Space Science (1993).

Astronaut Biography of Elena V. Kondakova

Personal Information:

Born 30 March 1957, in Mitischi, Moscow Region.

Elena is married to Valerii V. Ryumin, former astronaut and director of the Russian *Mir* program. They have one child.

Elena really enjoyed her trips to Mir. She once told Time magazine, "I personally consider it a vacation. You don't have to cook. You don't have to do laundry. You don't have to take care of the kids."

On 23 March 2001, she traveled with several other cosmonauts and private citizens to the South Pacific where they watched the *Mir* Space station re-enter Earth's atmosphere. It was a sad thing for her and many others to watch the end to the place they had called home for a time.

Elena enjoys the theater, river fishing, reading, and traveling.

Education:

Kondakova graduated from Moscow Bauman High Technical College in 1980.

Experience:

Upon graduation, Kondakova began to work at RSC-Energia completing science projects, experiments and research work.

Astronaut Experience:

In 1989, she was selected as a cosmonaut candidate by RSC-Energia Main Design Bureau and sent to Gagarin Cosmonaut Training Center to start general space training. After finishing the course in March 1990, Kondakova was qualified as "test cosmonaut".

Space Flight Experience:

From January through June of 1994, she was training for the "17th Main" mission and "Euromir-94" flight as flight engineer of the prime crew. Between 4 October 1994 and 22 March 1995, she served on the spacecraft "*Soyuz* TM-17" and the space station "Mir" as a flight engineer of the 17th main mission. She spent 169 days in space. The program included a month long joint flight with German Astronaut Ulf Merbold.

Most recently, she was a mission specialist on STS-84 (May 15-24, 1997), NASA's sixth Shuttle mission to rendezvous and dock with the Russian Space Station Mir. Mission duration was 9 days, 5 hours and 20 minutes. In completing her second flight, Kondakova had logged over 178 days in space.

Special Honors:

Elena Kondakova has been honored as a Hero of Russia.

Astronaut Biography of Eileen Marie Collins

Personal Information:

Born 19 November 1956, in Elmira, New York.

Eileen knew she wanted to be a pilot when she was ten and her father took her to see gliders fly. "I always thought that someday I would do that," she told reporters.

Eileen was the second of four children.

When she was 17 she started spending some of her savings from her sales jobs on flying lessons.

After High School, Eileen put herself through community college by working in a catalog showroom. She earned her pilot's license while still in college. And after college she enlisted in the Air Force and learned to fly jets.

In 1995, Eileen became the first woman to pilot a space shuttle. And in 1999 she became the first woman Shuttle Commander.

Eileen is married to Pat Youngs, a Delta Air lines pilot. They have a daughter Bridget who was three when she was on STS-93, her most recent flight. "My daughter thinks that all moms fly the space shuttle," She told Space.com.

Eileen enjoys running, golf, hiking, camping, reading, photography, and astronomy.

Education:

Eileen Collins graduated from Elmira Free Academy, Elmira, New York, in 1974. She received an Associate of Science degree in mathematics/science from Corning Community College in 1976, and a Bachelor of Arts degree in mathematics and economics from Syracuse University in 1978. In 1986 she earned a Master of Science degree in operations research from Stanford University, and later a Master of Arts degree in space systems management from Webster University in 1989.

Experience:

Collins graduated in 1979 from Air Force Undergraduate Pilot Training at Vance AFB, Oklahoma, where she was a T-38 instructor pilot until 1982.

From 1983 to 1985, she was a C-141 aircraft commander and instructor pilot at Travis AFB, California. She spent the following year as a student with the Air Force Institute of Technology.

From 1986 to 1989, she was assigned to the U.S. Air Force Academy in Colorado, where she was an assistant professor in mathematics and a T-41 instructor pilot.

She was selected for the astronaut program while attending the Air Force Test Pilot School at Edwards AFB, California, from which she graduated in 1990.

She has logged over 5,000 hours in 30 different types of aircraft.

Astronaut Experience:

Selected by NASA in January 1990, Colonel Collins became an astronaut in July 1991. Initially assigned to Orbiter Engineering Support, Collins has also served on the astronaut support team responsible for Orbiter pre-launch checkout, final launch configuration, crew ingress/egress, landing/recovery, worked in Mission Control as a spacecraft communicator (CAPCOM), served as the Astronaut Office Spacecraft Systems Branch Chief, Chief Information Officer, Shuttle Branch Chief, and Astronaut Safety Branch Chief.

A veteran of three space flights, Collins has logged over 537 hours in space. She served as pilot on STS-63 (February 3-11, 1995) and STS-84 (May 15-24, 1997), and was the commander on STS-93 (July 22-27, 1999). Collins will serve as Commander of STS-114 scheduled to launch in late 2002.

Space Flight Experience:

STS-63 (February 3-11, 1995) was the first flight of the new joint Russian-American Space Program. Mission highlights included the rendezvous with the Russian Space Station Mir, operation of Spacehab, the deployment and retrieval of an astronomy satellite, and a space walk. Collins was the first woman pilot of a space shuttle.

STS-84 (May 15-24, 1997) was NASA's sixth Shuttle mission to rendezvous and dock with the Russian Space Station Mir. During the flight, the crew conducted a number of secondary experiments and transferred nearly 4 tons of supplies and experiment equipment between *Atlantis* and the *Mir* station.

STS-93 *Columbia* (July 23-27, 1999) was the first Shuttle mission to be commanded by a woman. STS-93 highlighted the deployment of the Chandra X-Ray Observatory. Designed to conduct comprehensive studies of the universe, the telescope has enabled scientists to study exotic phenomena such as exploding stars, quasars, and black holes.

Organizations:

Colonel Collins is a member of the Air Force Association, Order of Daedalians, Women Military Aviators, U.S. Space Foundation, the American Institute of Aeronautics and Astronautics, and the Ninety-Nines.

Special Honors:

Defense Superior Service Medal, Distinguished Flying Cross, Defense Meritorious Service Medal, Air Force Meritorious Service Medal with one oak leaf cluster, Air Force Commendation Medal with one oak leaf cluster, Armed Forces Expeditionary Medal for service in Grenada (Operation Urgent Fury, October 1983), French Legion of Honor, NASA Outstanding Leadership Medal, NASA Space Flight Medals.

Astronaut Biography of Wendy B. Lawrence

Personal Information:

Born 2 July 1959, in Jacksonville, Florida.

Wendy has wanted to be an astronaut since she was a girl. "I sat there and watched Neil Armstrong walk on the moon for the very first time. I was 10 and I was just mesmerized by that event, and it became my childhood dream. I still remember saying to myself, 'that's what I want to do when I grow up. I want to be an astronaut.'"

She had the bug to go into space. "Often times I would sit down with my brother and we would watch Star Trek and watch Outer Limits." Wendy isn't as much of a science fiction fan these days. "I have seen some of the new Star Treks that have come out, I can't say I watch them as religiously. Of course some of the space movies they have out these days, I can't watch them because I say 'that's not technically correct, you can't do that'. But *Apollo* 13 was extremely well done."

"Not only did I want to be an Astronaut, but I grew up in a Navy family," She says. Her father was a Navy test pilot, so she grew up wanting to fly in the Navy too. When she was in junior high school, the Naval academy was finally opened to women. "Going to the Naval academy was very, very attractive to me because I knew many of the first several astronauts had gone to service academies and had been military aviators. So I thought that was a key stepping stone to take on the path of becoming both a pilot and an astronaut."

Wendy's dad was involved in the initial selection process of astronauts. "He had very good friends, people he had been a test pilot with, and flown in the Navy, like John Glenn and Alan Shepard who had become astronauts. So I didn't necessarily have to go to the astronauts, I could just go to my father, and get their advice through him."

Her mother was also a great help in becoming an astronaut, "I had the benefit, and still have the benefit, of having a mother who taught nursery school for 40 plus years. She has the rare gift of being able to absolutely mesmerize 3 and 4 year olds. All three of her children were the beneficiaries of that gift. She made learning fun."

Wendy enjoys running, rowing, and triathlons.

Wendy's advice is to, "be actively involved in making your dream come true. You have many resources that you may not realize. My teachers were a tremendous resource, they were a tremendous source of support, as was my family, my coaches were very helpful and supportive as well. The other thing I like to tell young people is they need to surround themselves with friends who will make them better people. They need to surround themselves with friends who will help them see their dream come true."

Education:

Wendy Lawrence graduated from Fort Hunt High School, Alexandria, Virginia in 1977. She received a Bachelor of Science degree in ocean engineering from U.S. Naval Academy in 1981, and a Master of Science degree in ocean engineering from Massachusetts Institute of Technology (MIT) and the Woods Hole Oceanographic Institution (WHOI) in 1988.

Experience:

Commander Lawrence graduated from the United States Naval Academy in 1981. A distinguished flight school graduate, she was designated as a naval aviator in July 1982. Lawrence has more than 1,500 hours flight time in six different types of helicopters and has made more than 800 shipboard landings.

While stationed at Helicopter Combat Support Squadron SIX (HC-6), she was one of the first two female helicopter pilots to make a long deployment to the Indian Ocean as part of a carrier battle group.

After completion of a master's degree program at MIT and WHOI in 1988, she was assigned to Helicopter Anti-Submarine Squadron Light THIRTY (HSL-30) as officer-in-charge of Detachment ALFA.

In October 1990, Lawrence reported to the U.S. Naval Academy where she served as a physics instructor and the novice women's crew coach.

Astronaut Experience:

NASA selected Commander Lawrence in March 1992. She reported to the Johnson Space Center in August 1992. She completed one year of training and is qualified for flight assignment as a Mission Specialist.

Her technical assignments within the Astronaut Office have included flight software verification in the Shuttle Avionics Integration Laboratory (SAIL) and Astronaut Office Assistant Training Officer.

After her first flight Commander Lawrence served as Director of Operations for NASA at the Gagarin Cosmonaut Training Center in Star City, Russia, with responsibility for the coordination and implementation of mission operations activities in the Moscow region for the joint U.S./Russian Shuttle/Mir program.

In September 1996 she began training for a 4-month mission on the Russian Space Station Mir, but in July 1997 NASA decided to replace Lawrence with her back up, Dr. David Wolf because Lawrence was too short to fit into the Russian space suits.

Because of her knowledge and experience with *Mir* systems and with crew transfer logistics, she flew with the crew of STS-86 (September 25 to October 6, 1997). She currently serves as the Astronaut Office representative for Space Station training.

"I'm the Astronaut Office representative to the training division," says Lawrence. "I also have the responsibilities for supporting the crews that are in training for missions on board the space station, the long-duration missions on the space station. So that keeps me pretty busy."

Space Flight Experience:

A veteran of three space flights, she has logged over 894 hours in space.

STS-67 Endeavour (March 2-18, 1995) was the second flight of the ASTRO observatory, a unique complement of three telescopes. During this 16-day mission, the crew conducted observations around the clock to study the far ultraviolet spectra of faint astronomical objects and the polarization of ultraviolet light coming from hot stars and distant galaxies. Mission duration was 399 hours and 9 minutes.

STS-86 *Atlantis* (September 25 to October 6, 1997) was the seventh mission to rendezvous and dock with the Russian Space Station Mir. Highlights included the delivery of a *Mir* attitude control computer, the exchange of U.S. crew members Mike Foale and David Wolf, a spacewalk by Scott Parazynski and Vladimir Titov to retrieve four experiments first deployed on *Mir* during the STS-76 docking mission, the transfer to *Mir* of 10,400 pounds of science and logistics, and the return of experiment hardware and results to Earth. Mission duration was 169 orbits in 259 hours and 21 minutes.

STS-91 *Discovery* (June 2-12, 1998) was the 9th and final Shuttle-Mir docking mission and marked the conclusion of the joint U.S./Russian Phase I Program. Mission duration was 235 hours, 54 minutes.

Organizations:

Commander Lawrence is a member of the Phi Kappa Phi, Association of Naval Aviation, Women Military Aviators, and Naval Helicopter Association.

Special Honors:

Commander Lawrence has been awarded the Defense Superior Service Medal, the Defense Meritorious Service Medal, the NASA Space Flight Medal, the Navy Commendation Medal and the Navy Achievement Medal. She is also the recipient of the National Navy League's Captain Winifred Collins Award for Inspirational Leadership (1986).

Astronaut Biography of Mary Ellen Weber

Personal Information:
Born 24 August 1962 in Cleveland, Ohio, but considers Bedford Heights, Ohio as her hometown.

Mary is married to Dr. Jerome Elkind, who is with Texas Instruments, Inc. and originally from Bayonne, New Jersey.

Mary is an avid skydiver. She has logged over 3,300 skydives since 1983. She received a silver medal in the U.S. National Skydiving Championships in the 20-person freefall formation event, in 1997, 1995, and 1991. She was also in the world's largest freefall formation in 1996, with 297 people. She told the Houston Chronicle, "There's just nothing like that exhilarating feeling."

Once Mary is assigned to a specific mission, she must give up her passion for skydiving for a while. NASA forbids any risky activity once astronauts are assigned to a mission.

Mary is also a golfer, an instrument rated pilot, and enjoys scuba diving.

Education:
Mary Weber graduated from Bedford High School in 1980. She received a Bachelor of Science degree in chemical engineering from Purdue University in 1984, and received a Ph.D. in physical chemistry from the University of California at Berkeley in 1988.

Experience:
During her undergraduate studies at Purdue, Dr. Weber was an engineering intern at Ohio Edison, Delco Electronics, and 3M. Following this, in her doctoral research at Berkeley, she explored the physics of gas-phase reactions involving silicon.

She then joined Texas instruments to research new processes for making computer chips. TI assigned her to a consortium of semiconductor companies, SEMATECH, and subsequently to Applied Materials, to create a revolutionary reactor for manufacturing next-generation chips.

She has received one patent and published eight papers in scientific journals.

Astronaut Experience:
NASA selected Dr. Weber for astronaut training in 1992.

In administrative assignments, Dr. Weber most recently worked with a venture capital firm to identify promising areas of space research and related companies for investment. In addition, she was the Legislative Affairs liaison at NASA Headquarters in Washington D.C., interfacing with Congress and traveling with the NASA Administrator. Prior to this appointment, she was Chairman of the procurement board for the Biotechnology Program contractor. Also, she served on a team designated to assess and revamp the Space Station research facilities.

Dr. Weber's principal technical assignments within the Astronaut Office have included Shuttle launch preparations at the Kennedy Space Center, payload and science development, and development of standards and methods for crew science training.

Space Flight Experience:

A veteran of two spaceflights, STS-70 and STS-101, Dr. Weber has logged over 450 hours in space.

STS-70 *Discovery* (July 13-22, 1995) was a mission which successfully delivered to orbit a critical NASA communications satellite, TDRS-G. Dr. Weber's primary responsibility was checking the systems of the satellite and sending it into its 22-thousand-mile orbit above the equator. She also performed biotechnology experiments, growing colon cancer tissues never before possible. She was the primary contingency spacewalk crewmember, and the medical officer. The STS-70 mission was completed in 142 orbits of the Earth, after traveling 3.7 million miles in 214 hours and 20 minutes.

STS-101 *Atlantis* (May 19-29, 2000) was the third Shuttle mission devoted to International Space Station construction. The crew repaired and installed a myriad of electrical and life-support components, both inside and out, and boosted the Station to a safe orbit. Dr. Weber's two primary responsibilities were flying the 60-foot robotic arm to maneuver a spacewalk crewmember along the Station surface, and directing the transfer of over three thousand pounds of equipment. The STS-101 mission was accomplished in 155 orbits of the Earth, after traveling 4.1 million miles in 236 hours and 9 minutes.

Astronaut Biography of Catherine G. "Cady" Coleman

Personal Information:

Born 14 December 1960, in Charleston, South Carolina.

Cady was a good student who was inspired by a great high school chemistry teacher, Mrs. Ruth Opps. "She passed her excitement on to me," Cady told a NASA reporter. "I discovered later that many things she talked about I didn't really understand at the time, but she made me want to know more."

As an undergraduate, in college Cady competed in intercollegiate athletics on the Massachusetts Institute of Technology's (MIT) crew team.

While at MIT, Cady was fascinated by polymers, which are long chains of molecules, similar to a string of spaghetti, but microscopic. Chemical reactions of polymers are interesting because the long chains react in strange ways. Cady says it's like when you make fudge. "Each time you would try to make it," she says, "you would put the same set of ingredients together. Yet, one time you get fudge that is all granular and crunchy, and the next it is a soupy mess that won't solidify. It all has to do with the processing — how you put the ingredients together, the baking time, the way you mixed it up. I always wanted to know how the differences could happen."

Cady's interest in chemistry had her headed for a career as a research scientist. But then something happened to change that. "Being an astronaut wasn't really a common occupation when I was a kid." Cady says. "I didn't think about it until I was in college, and Sally Ride came to speak at [my college]. Listening to her I thought, 'Wow, I want that job!' I wanted adventure in my life."

"Follow your heart and study whatever subject you have a passion for," Cady told NASA's Women in Microgravity web site. "When people tell you "No" or don't take you seriously - take it as an opportunity to accept a challenge. You may or may not change their minds, but in the process you will be doing your best to achieve your dreams. The sky is NOT the limit anymore!!!!"

Cady is married to glass artist Josh Simpson.

She enjoys flying, scuba diving, sports, and music.

Education:

Coleman graduated from W.T. Woodson High School, Fairfax, Virginia, in 1978. She received a Bachelor of Science degree in chemistry from the Massachusetts Institute of Technology in 1983, and a doctorate in polymer science and engineering from the University of Massachusetts in 1991.

Experience:

Lieutenant Colonel Coleman was commissioned as a 2nd lieutenant in the Air Force in 1983 and began graduate work at the University of Massachusetts. Her research focused on polymer synthesis using the olefin metathesis reaction, and polymer surface modification.

In 1988, Coleman entered active duty and was assigned to Wright-Patterson Air Force Base. As a research chemist at the Materials Directorate of the Wright Laboratory, she synthesized model compounds to investigate the use of organic polymers for third-order nonlinear optical applications such as advanced computers and data storage.

Coleman was a surface analysis consultant for the Long Duration Exposure Facility (launched from STS 41-C in 1984 and retrieved during STS-32 in 1990).

In addition to assigned military duties, Coleman was a volunteer test subject for the centrifuge program at the Crew Systems Directorate of the Armstrong Aeromedical Laboratory. She set several endurance and tolerance records during her participation in physiological and new equipment studies.

Astronaut Experience:

Lieutenant Colonel Coleman was selected by NASA in March 1992 and reported to the Johnson Space Center in August 1992. Initially assigned to the Astronaut Office Mission Support Branch and detailed to flight software verification in the Shuttle Avionics Integration Laboratory, Coleman later served as the special assistant to the Director of the Johnson Space Center. She served in the Astronaut Office Payloads and Habitability Branch, working with experiment designers to insure that payloads can be operated successfully in the microgravity environment of low earth orbit. As the lead astronaut for long term space flight habitability issues, Coleman led the effort to label the Russian segments of the International Space Station in English and also tracked issues such as acoustics and living accommodations aboard the station.

Space Flight Experience:

A veteran of two space missions, Coleman has logged over 500 hours in space. She was a mission specialist on STS-73, trained as a backup mission specialist for an injured crewmember on STS-83, and was lead mission specialist on STS-93. STS-73 *Columbia* (October 20 to November 5, 1995), was the second United States Microgravity Laboratory mission. The mission focused on materials science, biotechnology, combustion science, the physics of fluids, and numerous scientific experiments housed in the pressurized Spacelab module. In completing her first space flight, Coleman orbited the Earth 256 times, traveled over 6 million miles, and logged a total of 15 days, 21 hours, 52 minutes and 21 seconds in space. STS-93 *Columbia* (July 22-27, 1999) was a 5-day mission during which Coleman was the lead mission specialist for the deployment of the Chandra X-Ray Observatory. Designed to conduct comprehensive studies of the universe, the telescope will enable scientists to study exotic phenomena such as exploding stars, quasars, and black holes. Mission duration was 118 hours and 50 minutes.

Organizations:

Lieutenant Colonel Coleman is a member of the American Chemical Society, the Society for Photo-Optical Instrumentation Engineers (SPIE), the American Association of University Women, and the International Women's Air and Space Museum.

Astronaut Biography of Claudie André-Deshays Haigneré

Personal Information:

Born 13 May 1957 in Le Creusot, France.

Claudie was the first French woman astronaut.

She has a sister, Danielle, who is two years older than her, and a brother, Pascal, who is three years younger. Growing up, her father traveled all over the world on business and was often away from home. Claudie learned geography by following her dad's travels.

Claudie first became interested in space at age 12 when she saw the first moon landings. She was, "Discovering that dreams could become reality through the *Apollo* missions, then I read a lot about space and mainly manned space flight, leaving that in a secret part of my memory that reappeared in 1985 with the announcement for an astronaut's selection by the French space agency (CNES)."

She was interested in learning everything. "Exploration literature and TV or movies, documentaries about nature, animals and faraway regions and people. Also art history," she says. In school "everything was of interest for me: Math, physics, geography, history, life sciences, Latin, Greek, foreign languages, philosophy, and sports. I was very hardworking, because there was so much to learn."

Growing up, Claudie was interested in sports too. She says, "I was first interested in sports (gymnastics) to practice and teach, then quickly [became] interested in the body's movement physiology, and then medicine and rheumatology. Always on the same line with the body's performance and health."

She met fellow astronaut Jean-Pierre Haigneré while training for a mission in Russia. The two were married in Star City, Russia and have one daughter, Carla (now 4 years old). "She was with me during my training in Star City before my last flight, going with pleasure to the Russian kindergarten and speaking a quite perfect Russian." Before her 2001 launch, Claudie told France-2 television that she had packed, "Some music, some pictures, and I brought my daughter's little teddy bear."

She says her toughest challenge, as an astronaut, was to "learn the Russian language in one month, to be fluent not only for technical training but also to be culturally integrated in Star City."

"I have mainly 2 characteristics: curiosity and determination," she says explaining her success. Others can do as she did, "Learn more, look forward to make your project real; to give yourself a chance to become a doctor, to become an astronaut or whatever you choose."

Claudie enjoys contemporary art (paintings and sculptures), reading, and sports, especially gymnastics and golf.

"Space adventure is a fascinating reality," says Claudie. "I would like to see space development within ethical boundaries for the progress of ALL people. Space is a new dimension, a new tool, it requires a new perception of the risks and benefits analysis taking into consideration, economy, science, technology and human cultural values."

Claudie's goal for the space program, "End Goal: a human mission on Mars around 2025, and some steps before, maybe including a return to the Moon as a test bed for technology and life support for the Mars human mission. And a lot of robotic precursor missions to keep coherence between science, technological and human components of the endeavor. We will need private funding in addition to public funding. That means opening our activities to space commercialization but always with the respect of safety and ethical considerations, including the use of astronaut and space images."

She is working hard towards that goal. "We are preparing the next steps, a human presence on Mars for our kids. I am personally involved in the definition of the roadmap for this project inside the European space agency (Aurora Programme) and I try to get as much wide and active engagement in it by the general public, politicians and scientists."

Claudie's advice to others who want to be astronauts, "There is no recipe or magic formula. It is a long way, you must be highly motivated, available, open minded and every time ready to face a new challenge. Don't forget that you will need support from your family and your friends, so don't forget them on the way to space."

Education:

Graduation from Faculté de Médecine (PARIS-Cochin) and Faculté des Sciences (PARIS-VII). 1981: M. D. Rheumatologist, specialist in Aviation and Space Medicine. She holds certificates (Certificat d'Etudes Spécialisées -CES) in biology and sports medicine (1981), aviation and space medicine (1982), and rheumatology (1984). In 1986 she received a diploma (Diplôme d'Etudes Approfondies - DEA) in biomechanics and physiology. Her 1992 doctoral thesis was in Neurosciences.

Experience:

From 1984 to 1992, she worked at the Rheumatology Clinic and the Readaptation Service of the Cochin Hospital in Paris: research and application of diagnostic and therapeutic techniques in rheumatology and sports traumatology.

From 1985 to 1990, she also worked at the Neurosensorial Physiology Laboratory of CNRS in Paris. She was involved in development and preparation of scientific experiments in the field of human physiology, in particular with the experiments "Physalie" and "Viminal" integrated into the French-Soviet mission "Aragatz" to the *Mir* station in 1988, with Jean-Loup Chrétien aboard the *Mir* station. Her research topics were human adaptation of motor and cognitive systems in weightlessness.

Haigneré was selected as a French astronaut candidate in 1985 by CNES, the French space agency.

From 1990 – 1992, Haigneré was responsible for French and international space physiology and medicine programs within CNES's Life Sciences division in Paris. From 1989 to 1992, she was responsible for scientific coordination of the life sciences experiments aboard the French-Russian Antarès mission, which took place in 1992.

She regularly participated in parabolic flight missions aboard the French Zero-G training aircraft, the *Caravelle*.

In October 1992, she was assigned as the back-up cosmonaut to Jean-Pierre Haigneré for the French-Russian "Altair" mission from July 1-22, 1993. During this mission, she was responsible for monitoring the biomedical experiments as a member of the ground team at the Mission Control Center in Kaliningrad, near Moscow.

In September 1993, she began coordinating the scientific program of the 1996 French-Russian mission "Cassiopée", as well as for the French experiments aboard the ESA Mission EUROMIR '94.

In December 1994, Haigneré was assigned as research cosmonaut to the "Cassiopée" mission and began training in Star City near Moscow on 1 January 1995.

In 1997, she worked in Moscow as the French representative of the space-oriented, French-Russian company, STARSEM.

In May 1998, Haigneré was again selected as the back up for Jean-Pierre Haigneré, this time for the French-Russian mission "Perseus" to *Mir* in February 1999. She trained for an EVA and obtained the qualification of "cosmonaut engineer" both for the *Soyuz* vehicle and the *Mir* space station. During the mission, she was the crew interface coordinator at the Mission Control Center in Koroliev.

In July 1999 she received the qualification as *Soyuz* Return Commander. She was the first woman astronaut in this position, which qualifies her to command a three-person *Soyuz* capsule during its return from space.

On 1 November 1999, she joined the European Astronaut Corps whose home is the European Astronaut Center (EAC) in Cologne, Germany. She provided collateral duties to ESA Development projects for the European Microgravity Facilities for Columbus and supports the medical operational activities within the Directorate of Manned Spaceflight and Microgravity (MSM).

Space Flight Experience:

Haigneré was the first French woman to travel into space.

From 17 August to 2 September 1996 she participated in the Cassiopée mission launched on *Soyuz* TM-24. After docking with *Mir* she performed a great number of experiments in the field of Life Sciences (physiology and developmental biology), fluid physics and technology.

In January 2001, Haigneré took up training at Yuri A. Gagarin Training Center near Moscow for preparation as *Soyuz* board engineer. She flew on *Soyuz* taxi mission TM-33 on 21 October 2001. The main purpose of the mission was the exchange of the *Soyuz* spacecraft used as a crew escape vehicle. She and her Russian crew spent eight days on the International Space Station performing a number of scientific experiments selected by the French Space Agency CNES, including ESA and DLR experiments.

Organizations:

Honor Member of the Société Francaise de Médecine Aéronautique et Spatiale, Corresponding Member of the International Academy of Astronautics (IAA), Honor Member and Administrator of AAAF, Member of the ANAE (Académie de l'Air et de l'Espace)

Special Honors:

"Chevalier de la Légion d'Honneur" and "Chevalier de l'Ordre National du Mérite". She has received different steps of the Russian "Order of Friendship" in recognition of her long and successful involvement in French-Russian space cooperation, and the Russian "Medal of Personal Courage".

Astronaut Biography of Susan Kilrain

Personal Information:

Born 24 October 1961, in Augusta, Georgia.

Susan is a Naval aviator and test pilot. Only recently were these career paths open to women. "I am lucky to have been born in this time. Just in the nick of time, doors opened before me," Susan said at a military press conference.

As an astronaut, Susan gives presentations to various groups around the U.S. "Inside Business" reported on one of these events, "Part of what I'll try to do is instill a little bit of my love of science and engineering and how I used that to get into the astronaut program."

Susan is married to Colin James Kilrain of Braintree, Massachusetts.

She enjoys triathlons, martial arts, and playing the piano.

Education:

Susan graduated from Walnut Hill High School, Natick, Massachusetts, in 1979. She has a Master of Science degree in aerospace engineering from Georgia Institute of Technology, 1985.

Experience:

After graduating from undergraduate school, Susan worked as a Wind Tunnel Project Officer for Lockheed Corporation in Marietta, Georgia and earned her graduate degree.

She was commissioned in 1985 and designated a naval aviator in 1987.

Susan was selected to be a flight instructor in the TA-4J Skyhawk. She later flew EA-6A Electronic Intruders for Tactical Electronic Warfare Squadron 33 in Key West, Florida.

After completing Test Pilot School, she reported to Fighter Squadron 101 in Virginia Beach, Virginia for F-14 Tomcat training. She has logged over 3,000 flight hours in more than 30 different aircraft.

Astronaut Experience:

Commander Kilrain reported to the Johnson Space Center in March 1995. Following a year of training, she worked technical issues for the Vehicle Systems and Operations Branch of the Astronaut Office. She also served as spacecraft communicator (CAPCOM) in mission control during launch and entry for numerous missions.

She is currently assigned to the Office of Legislative Affairs at NASA Headquarters, Washington D.C.

Space Flight Experience:

A veteran of two space flights, Commander Kilrain has logged over 471 hours in space. She flew as pilot on STS-83 (April 4-8, 1997) and STS-94 (July 1-17, 1997).

Her first mission, STS-83 (April 4-8, 1997), was cut short because of problems with one of the Shuttle's three fuel cell power generation units. Mission duration was 95 hours and 12 minutes, traveling 1.5 million miles in 63 orbits of the Earth.

STS-94 (July 1-17, 1997) was a re-flight of the Microgravity Science Laboratory (MSL-1) Spacelab mission, and focused on materials and combustion science research in microgravity. Mission duration was 376 hours and 45 minutes, traveling 6.3 million miles in 251 orbits of the Earth.

Organizations:

Commander Kilrain is a member of Association of Naval Aviation, Association of Space Explorers, and The Georgia Tech Foundation.

Special Honors:

Commander Kilrain is a Distinguished Naval Graduate of Aviation Officer Candidate School, and a Distinguished Graduate of the United States Naval Test Pilot School, Class 103. She has been awarded the Defense Superior Service Medal, Navy Meritorious Service Medal, Navy Commendation Medal, Navy Achievement Medal, NASA Space Flight Medals (2), and the National Defense Service Medal.

Astronaut Biography of Kalpana Chawla

Personal Information:
Born 1 July 1961 in Karnal, India.

Kalpana is the first Indian woman astronaut.

She is from a market town north of Delhi in the state of Haryana in India. According to BBC News, her parents say she was an extremely intelligent and sometimes rebellious teenager. In 1984 she went to the United States to study aeronautical engineering.

"Kalpana was very tomboyish," said her mother in a recent interview with India's magazine, The Week. "She used to cut her own hair, never wore ironed clothes and learned karate."

Her teachers say she was an excellent student. One said, "Once she prepared a project on the environment in which she made huge, colorful charts and models depicting the sky and stars. Traces of her interest in space may perhaps be found in this streak."

Kalpana enjoys flying, hiking, backpacking, and reading. She holds Certificated Flight Instructor's license with airplane and glider ratings, Commercial Pilot's licenses for single- and multi-engine land and seaplanes, and Gliders, and instrument rating for airplanes. She enjoys flying aerobatics and tail-wheel airplanes.

"I like airplanes, it's that simple," she says. "The theoretical side is mentally challenging but flying for me is sheer fun. It appeals to all my senses. The astronaut's job requires a technical background and a strong desire to go out in the blue yonder."

Kalpana is a naturalized US citizen, and is married to Jean-Pierre Harrison, a flying instructor.

"When I joined engineering, there were only seven girls in the whole engineering college. I was the first girl to go into aerospace engineering," Kalpana remembers. "That's the message I want to give other women: do something because you really want to do it. Even if it is a goal which is not necessarily within reach."

Education:
Kalpana Chawla graduated from Tagore School, Karnal, India, in 1976. She earned a Bachelor of Science degree in aeronautical engineering from Punjab Engineering College, India in 1982, and a Master of Science degree in aerospace engineering from University of Texas in 1984. Her Doctorate of philosophy in aerospace engineering from University of Colorado was awarded in 1988.

Experience:
In 1988, Kalpana Chawla started work at NASA Ames Research Center in the area of powered-lift computational fluid dynamics. Her research concentrated on simulation of complex airflows encountered around aircraft such as the Harrier in "ground-effect."

Following completion of this project she supported research in mapping of flow solvers to parallel computers, and testing of these solvers by carrying out powered lift computations.

In 1993 Kalpana Chawla joined Overset Methods Inc., Los Altos, California, as Vice President and Research Scientist to form a team with other researchers specializing in simulation of moving multiple body problems. She was responsible for development and implementation of efficient techniques to perform aerodynamic optimization. Results of various projects that Kalpana Chawla participated in are documented in technical conference papers and journals.

Astronaut Experience:

Selected by NASA in December 1994, Kalpana Chawla reported to the Johnson Space Center in March 1995 as an astronaut candidate in the 15th Group of Astronauts.

After completing a year of training and evaluation, she was assigned as crew representative to work technical issues for the Astronaut Office EVA/Robotics and Computer Branches. Her assignments included work on development of Robotic Situational Awareness Displays and testing space shuttle control software in the Shuttle Avionics Integration Laboratory.

In January 1998, Dr. Chawla was assigned as crew representative for shuttle and station flight crew equipment. Subsequently, she was assigned as the lead for Astronaut Office's Crew Systems and Habitability section.

Space Flight Experience:

In November 1996, Dr. Chawla was assigned as mission specialist and prime robotic arm operator on STS-87 (November 19 to December 5, 1997). STS-87 was the fourth U.S Microgravity Payload flight and focused on experiments designed to study how the weightless environment of space affects various physical processes, and on observations of the Sun's outer atmospheric layers. Two members of the crew performed an EVA (spacewalk), which featured the manual capture of a Spartan satellite, in addition to testing EVA tools and procedures for future Space Station assembly. In completing her first mission, Kalpana Chawla traveled 6.5 million miles in 252 orbits of the Earth and logged 376 hours and 34 minutes in space.

She is currently assigned to the crew of STS-107 scheduled for launch 19 July 2002 on the shuttle *Columbia*. The research mission, which will look at combustion in space, as well as fluid experiments, and several biological experiments.

Astronaut Biography of Kathryn P. (Kay) Hire

Personal Information:

Born 26 August 1959 in Mobile, Alabama

As a high school freshman, Kay's teachers talked her into taking a junior engineering aptitude test because of her interest in science and math. When she did well, universities started sending her catalogues.

At first Kay wanted to go to school at Georgia Tech because of its great engineering program. When she met with a Navy ROTC representative about getting a scholarship she found out about the US Naval academy in Annapolis, Maryland. She became interested, and because of her good grades and school activities (tennis and yearbook) received a sponsorship from her congressman.

Kay didn't have the 20/20 vision necessary to become a pilot, and she really wanted to fly. Instead, she became one of the first women Naval Flight Officers. In the book, Cool Careers for Girls, she explains, "The Navy had had women pilots since 1973, but they had just opened the Naval Flight Officer position, the 'back-seaters' to women the year before I graduated."

She enjoys competitive sailing, snow skiing, scuba diving, and fishing.

"Be prepared to take advantage of unexpected opportunities," says Kay. "You never know what careers will open up in the future, especially in the field of space exploration."

Education:

Kay graduated from Murphy High School, Mobile, Alabama, in 1977. She has a Bachelor of Science degree in engineering and management from the U.S. Naval Academy, 1981, and a Master of Science degree in space technology from Florida Institute of Technology, 1991.

Experience:

After earning her Naval Flight Officer Wings in October 1982, Hire conducted worldwide airborne oceanographic research missions with Oceanographic Development Squadron Eight (VXN-8) based at NAS Patuxent River, Maryland. She flew as Oceanographic Project Coordinator, Mission Commander and Detachment Officer-in-Charge on board the specially configured P-3 aircraft.

Hire instructed student naval flight officers in the classroom, simulator, and on board the T-43 aircraft at the Naval Air Training Unit on Mather Air Force Base, California. She progressed from Navigation Instructor through Course Manager to Curriculum Manager and was awarded the Air Force Master of Flying Instruction.

In January 1989, Hire joined the Naval Air Reserve at NAS Jacksonville, Florida. Her tours of duty included Squadron Augment Unit VP-0545 and Anti-submarine Warfare Operations Center 0574 and 0374.

Hire became the first female in the United States assigned to a combat aircrew when she reported to Patrol Squadron Sixty-Two (VP-62) on May 13, She says, "I always felt the combat restrictions would go away, and eventually they did. When that happened I was right there ready to take advantage of the new opportunities."

As a Patrol Plane Navigator/Communicator she deployed to Iceland, Puerto Rico and Panama. Hire later served with CV-63 USS *Kitty Hawk* 0482, Tactical Support Center 0682, and Commander Seventh Fleet Detachment 111. Presently she is a member of U.S. Naval Central Command Detachment 108.

Astronaut Experience:

Hire began work at the Kennedy Space Center in May 1989, first as an Orbiter Processing Facility 3 Activation Engineer and later as a Space Shuttle Orbiter Mechanical Systems Engineer for Lockheed Space Operations Company. In 1991 she certified as a Space Shuttle Test Project Engineer (TPE) and headed the checkout of the Extravehicular Mobility Units (spacesuits) and Russian Orbiter Docking System. Hire was assigned Supervisor of Space Shuttle Orbiter Mechanisms and Launch Pad Swing Arms in 1994.

Selected as an astronaut by NASA in December 1994, Hire reported to the Johnson Space Center in March 1995.

After a year of training, she worked in mission control as a spacecraft communicator (CAPCOM). Hire flew as Mission Specialist-2 on STS-90 Neurolab (April 17 to May 3, 1998). During the 16-day Spacelab flight the seven member crew aboard Space Shuttle *Columbia* served as both experiment subjects and operators for 26 life science experiments focusing on the effects of microgravity on the brain and nervous system. The STS-90 flight orbited the Earth 256 times, covered 6.3 million miles, and logged over 381 hours in space. Hire currently serves as the Astronaut Office Lead for Shuttle Avionics Integration Lab (SAIL), Shuttle Payloads, and Flight Crew Equipment.

Organizations:

Commander Hire is a member of the Association of Naval Aviation, American Institute of Aeronautics and Astronautics, Institute of Navigation, and Society of Women Engineers, U.S. Sailing Association.

Special Honors:

Defense Superior Service Medal, National Defense Service Medal, Armed Forces Reserve Medal, Coast Guard Special Operations Service Ribbon, Navy and Marine Corps Overseas Service ribbon, Space Coast Society of Women Engineers Distinguished New Woman Engineer for 1993.

Astronaut Biography of Janet Lynn Kavandi

Personal Information:
Born 17 July 1959 in Springfield, Missouri.

"I was five or six when I first became interested in space," says Janet. "I lived on a farm in Missouri when I was very young, so the evening skies were filled with stars. My father and I would sit outside and talk about the first space flights, and whether or not one of the satellites passing overhead might contain one of the first astronauts. We wondered aloud what it would be like to be up there looking down at the Earth. My interest in space grew as I got older."

Janet's favorite studies in school were math and science, especially astronomy. She did a lot of reading on her own about space.

"I first became seriously interested in becoming an astronaut after the first females were selected to the Space Shuttle program in the late 1970's. When I saw the IMAX film, 'The Dream is Alive', I was even more inspired to become an astronaut. I put in my application to NASA in 1986, and updated it each year until I was finally interviewed in 1994." What Janet likes most about being an astronaut is the view in space. "The obvious primary benefit of this job is the privilege of flying in space. We have the rare opportunity to view our planet from a location that few people will have the ability to visit. I especially enjoy looking at the earth and stars during the night passes. The lightning storms are an incredible sight from space."

Janet is married to John Kavandi. They have two children.

She enjoys snow skiing, hiking, camping, horseback riding, windsurfing, flying, scuba diving, and piano.

"I believe in the exploration of space for the sake of basic knowledge and for the potential to set up colonies on the Moon and Mars. We still have a tremendous amount to learn about our solar system and the universe. We are just at the beginning of our understandings of the physics of space. Since our species has almost completely populated almost every continent on Earth, we will most likely need to either severely restrict population growth, or expand to neighboring planets. This expansion will be difficult, and will take many years to achieve. Our experience on board the Mir and the International Space Station will help further our understanding of the effects of microgravity and space radiation on the human body. We will learn what needs to be done to support life on a remote outpost."

Education:
Janet Kavandi graduated from Carthage Senior High School, Carthage Missouri, in 1977. She graduated Magna Cum Laude with a Bachelor of Science Degree in Chemistry from Missouri Southern State College, Joplin in 1980, and a Master of Science Degree in Chemistry from the University of Missouri, Rolla in 1982. Her Doctorate was awarded in 1990 in Analytical Chemistry from the University of Washington, Seattle.

Experience:
Following graduation in 1982, Dr. Kavandi accepted a position at Eagle-Picher Industries in Joplin, Missouri, as an engineer in new battery development for defense applications. In 1984, she accepted a position as an engineer in the Power Systems Technology Department of the Boeing Aerospace Company in Seattle, Washington. During her ten years at Boeing, Kavandi supported numerous programs, proposals and "red teams" in the energy storage systems area. She was

lead engineer of secondary power for the Short Range Attack Missile II, and principal technical staff representative involved in the design and development of thermal batteries for Sea Lance and the Lightweight Exo-Atmospheric Projectile. Other programs she supported include Space Station, Lunar and Mars Base studies, Inertial Upper Stage, Advanced Orbital Transfer Vehicle, Get-Away Specials, Air Launched Cruise Missile, Minuteman, and Peacekeeper. In 1986, while still working for Boeing, she was accepted to graduate school at the University of Washington, where she began working toward her doctorate in analytical chemistry. Her doctoral dissertation involved the development of a pressure-indicating coating that uses oxygen quenching of porphyrin photoluminescence to provide continuous surface pressure maps of aerodynamic test models in wind tunnels. Her work on pressure indicating paints has resulted in two patents. Dr. Kavandi has also published and presented several papers at technical conferences and in scientific journals.

Dr. Kavandi was selected as an astronaut candidate by NASA in December 1994 and reported to the Johnson Space Center in March 1995. Following an initial year of training, she was assigned to the Payloads and Habitability Branch where she supported payload integration for the International Space Station. Between her second and third flights (STS-99 and STS-104) Dr. Kavandi worked in the Robotics Branch where she trained on both the shuttle and space station robotic manipulator systems.

Space Flight Experience:

A three-flight veteran, Dr. Kavandi has logged over 33 days in space, traveling over 13.1 million miles in 535 Earth orbits. Dr. Kavandi served as a mission specialist on STS-91 (June 2-12, 1998) the 9th and final Shuttle-Mir docking mission, concluding the joint U.S./Russian Phase 1 program. Following the mission she worked as a CAPCOM (spacecraft communicator) in NASA's Mission Control Center. On her second mission, she served aboard STS-99 (February 11-22, 2000), the Shuttle Radar Topography Mission, which mapped more than 47 million miles of the Earth's land surface to provide data for a highly accurate three-dimensional topographical map. Most recently, she served aboard STS-104/ISS Assembly Flight 7A (July 12-24, 2001) on the 10th mission to the International Space Station. The shuttle crew installed the joint airlock "Quest" and conducted joint operations with the Expedition-2 crew.

Special Honors:

Dr. Kavandi was elected to Who's Who Among Students in American Universities and Colleges, 1980; Who's Who of Emerging Leaders in America, 1989-90, 1991-92; and Who's Who in the West, 1987-88. She was awarded certificates for Team Excellence and Performance Excellence from Boeing Missile Systems in 1991. She is also the recipient of three NASA Space Flight Medals for shuttle flights STS-91, STS-99, and STS-104. She was presented the NASA Exceptional Service Medal in 2001.

Astronaut Biography of Julie Payette

Personal Information:

Born 20 October 1963 in Montreal, Quebec, Canada.

Julie is currently the Chief Astronaut for the Canadian Space Agency. "When I was young I watched TV and saw astronauts in their space suits driving the Moon buggy," Julie told an audience at Canada's National Engineering Week 2001. She explained why she wanted to be an astronaut. "And I was saying wow, that's so great, I'd love to do that. I was so inspired, that's what I wanted to do. And one of the best disciplines to prepare someone to do that kind of work is engineering."

When asked what her favorite food was while on her space mission she said, "I liked the vegetables, I liked the pasta dishes. So I don't think I have a favorite. Well, maybe the Smarties."

Julie enjoys running, skiing, racquet sports and scuba diving. She holds a multi-engine commercial pilot license with instrument and float ratings.

Her advice to young people, "Study science, even if that's not the kind of career you're going to do one day, because it will give you the methodology, the way of thinking, the options, and tools to do anything you want to do."

Julie plays piano and has sung with the Montreal Symphonic Orchestra Chamber Choir, the Piacere Vocale in Basel, Switzerland, and with the Tafelmusik Baroque Orchestra Choir in Toronto, Canada.

She is fluent in French and English, and conversational in Spanish, Italian, and Russian.

Education:

Julia Payette attended primary and secondary school in Montreal, Quebec. In 1980 she received one of only six available Canadian scholarships to attend the International United World International College (UWC) of the Atlantic in South Wales, UK (1980). She graduated with an International Baccalaureate in 1982. She then went on to get a Bachelor of Science in Electrical Engineering in 1986 from McGill University, Montreal and a Master of Applied Science in Computer Engineering in 1990 from the University of Toronto.

Experience:

Before joining the Canadian space program, Ms. Payette was a systems engineer for IBM Canada conducting research in computer systems, natural language processing, automatic speech recognition and the application of interactive technology to space. In 1991 she was a visiting scientist to the IBM Research Laboratory, Zurich, Switzerland. And in 1992 a Research engineer at the Speech Research Group, Bell-Northern Research / Nortel, Montreal, Canada.

Astronaut Experience:

Captain Payette was selected by the Canadian Space Agency (CSA) as one of four astronauts amongst a field of 5,330 applicants in June 1992. After undergoing basic training in Canada, she worked as a technical advisor for the MSS (Mobile Servicing System), an advanced robotics system and Canada's contribution to the International Space Station. In 1993, Captain Payette established the Human-Computer Interaction (HCI) Group at the Canadian Astronaut Program and served as a technical specialist on the NATO International Research Study Group (RSG-10) on speech processing (1993-1996). In preparation for a space assignment, Captain Payette obtained her commercial pilot license, studied Russian and logged over 120 hours of reduced gravity flight time aboard various microgravity aircraft. In April 1996, Captain Payette completed a deep-sea diving hard suit-training program in Vancouver, British *Columbia* and was certified as a one-atmosphere diving suit operator. Captain Payette obtained her captaincy on the CT-114 military jet at the Canadian Air Force Base in Moose Jaw, Saskatchewan in February 1996. She obtained her military instrument rating in 1997 and continues to fly with the training squadron whenever possible. Captain Payette has logged more than 800 hours of flight time, including 450 hours on high performance jet aircraft.

Space Flight Experience:

Captain Payette reported to the NASA Johnson Space Center in Houston, Texas in August 1996. She completed initial astronaut training in April 1998 and was assigned to work technical issues in robotics for the Astronaut Office. Julie Payette flew on Space Shuttle *Discovery* from May 27 to June 6, 1999 as part of the crew of STS-96. During the mission, the crew performed the first manual docking of the Shuttle to the International Space Station (ISS), and delivered 4 tons of logistics and supplies to the Station. On *Discovery*, Ms. Payette served as a mission specialist and was only the third Canadian astronaut to operate the Canadarm on orbit. The STS-96 mission was accomplished in 153 orbits of the Earth, traveling 4 million miles in 9 days, 19 hours and 13 minutes. Ms. Payette became the first Canadian to participate in an ISS assembly mission and to board the Space Station. Ms. Payette is the Chief Astronaut for the Canadian Space Agency. She currently works as a member of the Crew Test Support Team in support of ISS activities in Russia and in Europe. She divides her time between this work and astronaut currency training in Houston.

Organizations:

Julie Payette is a Member of l'Ordre des Ingénieurs du Québec and was appointed a member of the Natural Sciences and Engineering Research Council of Canada (NSERC). She is also a Fellow of the Canadian Academy of Engineering.

Special Honors:

Captain Payette has been awarded a number of scholarships, honorary degrees, and educational awards including the Canadian Council of Professional Engineers 1994 distinction for exceptional achievement by a young engineer She has also received the Chevalier de l'Ordre de la Pléiade de la francophonie (2000), and the National Order of Québec (2000).

Astronaut Biography of Pamela Ann Melroy

Personal Information:

Born 17 September 1961, in Palo Alto, California. Considers Rochester, New York, to be her hometown.

When Pam was seven and on vacation with her parents, David and Helen Melroy, they stopped to gather around a television to watch as *Apollo* 11 astronauts Neil Armstrong and Buzz Aldrin stepped on to the lunar surface for the first time. "This seemed like the most important thing anybody could do," she told the Houston Chronicle. That day she decided to be an astronaut.

In the *Apollo* days, NASA's astronauts were all male and almost all were military test pilots. "I never knew anyone who was a pilot. I didn't know anything about it," Pam recalled. "I just said, 'OK. All right. I will just have to go and become a military test pilot.'"

And that's just what she did. "I definitely had no idea what I was getting into," Pam said. "I graduated from college, went to graduate school and showed up at Air Force jet pilot school, never having flown an airplane before — throw me in the deep end. I loved it. It was a blast."

Before her first flight she said, "It's obviously a big deal to my mom, and it's exciting to me." Pam hopes more women will get involved in the space program. "I think the doors are really starting to open, and it won't be a big deal in a few years. *But for now, if someone looks at me and says, 'If she can do it, I can do it,' then I'm happy."*

Pam is married to Christopher Wallace.

She enjoys theatre, tap and jazz dancing, reading, cooking, and flying.

Education:

Melroy graduated from Bishop Kearney High School, Rochester, New York, in 1979. She has a Bachelor of Science degree in physics and astronomy from Wellesley College in 1983, and a Master of Science degree in earth & planetary sciences from Massachusetts Institute of Technology, 1984.

Experience:

Melroy was commissioned through the Air Force ROTC program in 1983. After completing her Master's degree, she attended Undergraduate Pilot Training at Reese Air Force Base in Lubbock, Texas and graduated in 1985. She flew the KC-10 for six years at Barksdale Air Force Base in Bossier City, Louisiana, as a copilot, aircraft commander and instructor pilot. Melroy is a veteran of JUST CAUSE and DESERT SHIELD/DESERT STORM, with over 200 combat and combat support hours. In June 1991, she attended the Air Force Test Pilot School at Edwards Air Force Base. Upon her graduation, she was assigned to the C-17 Combined Test Force, where she served as a test pilot until her selection for the astronaut program. She has logged over 5,000 hours flight time in over 45 different aircraft.

Astronaut Experience:

Selected as an astronaut candidate by NASA in December 1994, Melroy reported to the Johnson Space Center in March 1995. She completed a year of training and evaluation and is qualified for flight assignment as a space shuttle pilot. Initially assigned to astronaut support duties for launch and landing, she has also worked Advanced Projects for the Astronaut Office. Melroy is assigned as one of two pilots on STS-112, scheduled to launch in 2002.

Space Flight Experience:

Lieutenant Colonel Melroy was a pilot on STS-92 in 2000 and has logged over 309 hours in space.

STS-92 *Discovery* (October 11-24, 2000) was launched from the Kennedy Space Center, Florida and returned to land at Edwards Air Force Base, California. During the 13-day flight, the seven-member crew attached the Z1 Truss and Pressurized Mating Adapter 3 to the International Space Station using *Discovery's* robotic arm and performed four space walks to configure these elements. This expansion of the ISS opened the door for future assembly missions and prepared the station for its first resident crew.

Melroy also operated a 3D IMAX movie camera during the mission to collect footage for the IMAX movie "Space Station 3D" about the assembly of the ISS.

The STS-92 mission was accomplished in 202 orbits, traveling 5.3 million miles in 12 days, 21 hours, 40 minutes and 25 seconds.

Organizations:

Lieutenant Colonel Melroy is a member of the Society of Experimental Test Pilots, the Order of Daedalians, and the Ninety-Nines.

Special Honors:

Lieutenant Colonel Melroy is the recipient of the Air Force Meritorious Service Medal, First Oak Leaf Cluster; Air Medal, First Oak Leaf Cluster; Aerial Achievement Medal, First Oak Leaf Cluster; and Expeditionary Medal, First Oak Leaf Cluster.

Astronaut Biography of Peggy A. Whitson

Personal Information:

Born 9 February 1960 in Mt. Ayr, Iowa, but her hometown is Beaconsfield, Iowa.

Peggy is co-inventor of a blood serum sampling device that does not require a freezer to preserve the sample. It was invented for space use, but has great application for field workers in remote areas like Africa where freezers aren't available.

She enjoys windsurfing, biking, basketball, and water skiing.

Peggy is married to Clarence F. Sams.

Education:

Peggy Whitson graduated from Mt. Ayr Community High School, Mt. Ayr, Iowa, in 1978. She received a Bachelor of Science degree in biology/chemistry from Iowa Wesleyan College in 1981, and a doctorate in biochemistry from Rice University in 1985.

Experience:

From 1981 to 1985, Whitson conducted her graduate work in biochemistry at Rice University, Houston, Texas, as a Robert A. Welch Pre-doctoral Fellow. Following completion of her graduate work she continued at Rice University as a Robert A Welch Postdoctoral Fellow until October 1986.

Following this position, she began her studies at NASA Johnson Space Center, Houston, Texas, as a National Research Council Resident Research Associate. From April 1988 until September 1989, Whitson served as the Supervisor for the Biochemistry Research Group at KRUG International, a medical sciences contractor at NASA-JSC.

From 1991-1997, Whitson was an Adjunct Assistant Professor in the Department of Internal Medicine and Department of Human Biological Chemistry and Genetics at University of Texas Medical Branch, Galveston, Texas.

In 1997, Whitson began a position as Adjunct Assistant Professor at Rice University in the Maybee Laboratory for Biochemical and Genetic Engineering.

Astronaut Experience:

From 1989 to 1993, Whitson worked as a Research Biochemist in the Biomedical Operations and Research Branch at NASA-JSC. In 1990, she gained the additional duties of Research Advisor for the National Research Council Resident Research Associate. From 1991-1993, she served as Technical Monitor of the Biochemistry Research Laboratories in the Biomedical Operations and Research Branch. From 1991-1992 she was the Payload Element Developer for Bone Cell Research Experiment (E10) aboard SL-J (STS-47), and was a member of the US-USSR Joint Working Group in Space Medicine and Biology.

In 1992, she was named the Project Scientist of the Shuttle-Mir Program (STS-60, STS-63, STS-71, *Mir* 18, *Mir* 19) and served in this capacity until the conclusion of the Phase 1A Program in 1995. From 1993-1996 Whitson held the additional responsibilities of the Deputy Division Chief of the Medical Sciences Division at NASA-JSC. From 1995-1996 she served as Co-Chair of the U.S.-Russian Mission Science Working Group.

In April 1996, she was selected as an astronaut candidate and started training, in August 1996. Upon completing two years of training and evaluation, she was assigned technical duties in the Astronaut Office Operations Planning Branch and served as the lead for the Crew Test Support Team in Russia from 1998-99.

Space Flight Experience:

Whitson is the second woman resident of the ISS. She will serve as a prime crewmember on Increment 5, which will launch aboard Utilization Flight 2 (STS-111) from Florida. STS-111 is scheduled to launch in May 2002 on the shuttle Endeavour. It will be the 14th flight to the international space station. The Shuttle will carry a Multi-Purpose Logistics Module and a Mobile Servicing System. The Multipurpose Logistics Module, or MPLM, carries experiment racks and three stowage and resupply racks to the station. The Mobile Base System is installed on the Mobile Transporter to complete the Canadian Mobile Servicing System, or MSS. The mechanical arm will have the capability to "inchworm" from the U.S. Lab fixture to the MSS and travel along the Truss to work sites.

Whitson and the rest of the crew is scheduled to stay on the ISS until the Space Shuttle Endeavour returns on STS-113, currently scheduled for October 2002. During their four months in space, the crew will conduct experiments in human physiology, microgravity and space product development.

Special Honors:

Dr. Whitson has received a Group Achievement Award for Shuttle-Mir Program (1996), American Astronautical Society Randolph Lovelace II Award (1995), NASA Tech Brief Award (1995), NASA Space Act Board Award (1995, 1998), NASA Silver Snoopy Award (1995), NASA Exceptional Service Medal (1995), NASA Certificate of Commendation (1994), NASA Sustained Superior Performance Award (1990), and a Krug International Merit Award (1989).

Astronaut Biography of Laurel Blair Salton Clark

Personal Information:

Born 10 March 1961 in Ames, Iowa, but considers Racine, Wisconsin, to be her hometown.

Laurel is married to Jonathan B. Clark (Captain, USN). They have one child.

Her mother and stepfather, Dr. and Mrs. R.J.C. Brown, reside in Racine, Wisconsin. Her father and stepmother, Mr. and Mrs. Robert Salton, reside in Albuquerque, New Mexico. His parents, Colonel (ret) & Mrs. E.B. Clark III, reside in Alexandria, Virginia.

She enjoys scuba diving, hiking, camping, biking, parachuting, flying, and traveling.

Education:

Laurel Blair Salton Clark graduated from William Horlick High School, Racine Wisconsin in 1979. She received a Bachelor of Science degree in zoology from the University of Wisconsin-Madison in 1983 and doctorate in medicine from the same school in 1987.

Experience:

During medical school Commander Dr. Laurel Blair Salton Clark did active duty training with the Diving Medicine Department at the Naval Experimental Diving Unit in March 1987. After completing medical school, Dr. Clark underwent postgraduate Medical education in Pediatrics from 1987-1988 at Naval Hospital Bethesda, Maryland.

The following year Dr. Clark completed Navy undersea medical officer training at the Naval Undersea Medical Institute in Groton Connecticut and diving medical officer training at the Naval Diving and Salvage Training Center in Panama City, Florida, and was designated a Radiation Health Officer and Undersea Medical Officer.

She was then assigned as the Submarine Squadron Fourteen Medical Department Head in Holy Loch Scotland. During that assignment she dove with U.S. Navy divers and Naval Special Warfare Unit Two Seals and performed numerous medical evacuations from US submarines.

After two years of operational experience she was designated as a Naval Submarine Medical Officer and Diving Medical Officer. She underwent 6 months of aeromedical training at the Naval Aerospace Medical Institute in Pensacola, Florida and was designated as a Naval Flight Surgeon.

She was stationed at MCAS Yuma, Arizona and assigned as Flight Surgeon for a Marine Corps AV-8B Night Attack Harrier Squadron (VMA 211). She made numerous deployments, including one overseas to the Western Pacific, and practiced medicine in austere environments, and flew on multiple aircraft. Her squadron won the Marine Attack Squadron of the year for its successful deployment. She was then assigned as the Group Flight Surgeon for the Marine Aircraft Group (MAG 13). Prior to her selection as an astronaut candidate she served as a Flight Surgeon for the Naval Flight Officer advanced training squadron (VT-86) in Pensacola, Florida. Commander Clark is Board Certified by the National Board of Medical Examiners and holds a Wisconsin Medical License. Her military qualifications include Radiation Health Officer, Undersea Medical Officer, Diving Medical Officer, Submarine Medical Officer, and Naval Flight Surgeon. She is a Basic Life Support Instructor, Advanced Cardiac Life Support Provider, Advanced Trauma Life Support Provider, and Hyperbaric Chamber Advisor.

Astronaut Experience:

Selected by NASA in April 1996, Dr. Clark reported to the Johnson Space Center in August 1996. After completing two years of training and evaluation, she was qualified for flight assignment as a mission specialist.

From July 1997 to August 2000 Dr. Clark worked in the Astronaut Office Payloads/Habitability Branch.

Space Flight Experience:

She is currently assigned to the crew of STS-107 scheduled for launch in July 2002. The mission is an Extended Duration Orbiter research mission expected to last 13 to 16 days. The primary payload is the Spacehab Double Module. This is the first flight of the SHI Research Double Module (SHI/RDM). It will also fly the Fast Reaction Experiments Enabling Science, Technology, Applications and Research (FREESTAR) module.

Organizations:

Aerospace Medical Association, Society of U.S. Naval Flight Surgeons.

Special Honors:

Navy Commendation Medals (3); National Defense Medal, and Overseas Service Ribbon.

Astronaut Biography of Sandra H. Magnus

Personal Information:

Born 30 October 1964 in Belleville, Illinois.

Sandra wanted to become an astronaut when she was a student in high school and she made a plan to become an astronaut. There were detours to her plan, but even those detours helped her achieve her ultimate goal.

Sandra still has dreams. She told the St. Louis Post-Dispatch, "I would eventually like to be a space station resident for a while."

Sandra urges students to set goals and do their best. "The path you end up on may not be what you planned, but you have nothing to lose if you do your best," she says.

She enjoys soccer, reading, travel, and water activities.

Education:

Sandra Magnus graduated from Belleville West High School, Belleville, Illinois, in 1982. She received a bachelor degree in physics and a masters degree in electrical engineering from the University of Missouri-Rolla in 1986 and 1990, respectively, and a doctorate from the School of Material Science and Engineering at the Georgia Institute of Technology in 1996.

Experience:

During 1986 to 1991, Magnus worked for McDonnell Douglas Aircraft Company as a stealth engineer where she worked on internal research and development studying the effectiveness of RADAR signature reduction techniques. She was also assigned to the Navy's A-12 Attack Aircraft program primarily working on the propulsion system until the program was cancelled.

From 1991 to 1996, Magnus completed her thesis work, which was supported by NASA-Lewis Research Center through a Graduate Student Fellowship and involved investigations on materials of interest for "Scandate" thermionic cathodes. Thermodynamic equilibria studies along with conductivity and emission measurements on compounds in the Ba

O.SC2O3.WO3 ternary system were conducted to identify compounds with potential use in these types of cathodes.

Astronaut Experience:

Dr. Magnus was selected by NASA in April 1996. She reported to the Johnson Space Center in August 1996. Having completed two years of training and evaluation, she is qualified for flight assignment as a Mission Specialist.

From January 1997 through May 1998 Dr. Magnus worked in the Astronaut Office Payloads/Habitability Branch. Her duties involved working with ESA, NASDA and Brazil on science freezers, glove boxes and other facility type payloads.

In May 1998 Dr. Magnus was assigned as a "Russian Crusader" which involves travel to Russia in support of hardware testing and operational products development.

Space Flight Experience:

Currently, Dr. Magnus is assigned to STS-112, *Atlantis*, scheduled to launch in August 2002. The mission will be the Fifteenth Shuttle flight to the ISS. Primary payload is the right-side truss segment. Magnus will operate the shuttle's robotic arm to extract the truss from the shuttle's cargo bay and install it across the top of the space station. Two fellow crewmembers will perform an EVA to attach solar panels and other equipment to the truss.

One of her other mission duties is stowage and transfer. "I'm like the load master," she said. She is responsible for ensuring the proper balance of the spacecraft and that everything is in its proper place for launch and re-entry.

Organizations:

Dr. Magnus is a member to ASM/TMS (Metallurgical/Material Society), and the Material Research Society.

Special Honors:

Dr. Magnus has received the Outstanding Graduate Teaching Assistant Award (1994 and 1996), Saturn Team Award (1994), and Performance Bonus Award (1989).

Astronaut Biography of Heidemarie M. Stefanyshyn-Piper

Personal Information:

Born 7 February 1963 in St. Paul, Minnesota.

Heidi became interested in airplanes when she was four. She was five when she became interested in space after watching the *Apollo* 11 Moon landing. During the *Apollo-Soyuz* mission she remembers thinking, "I could be one of the astronauts, and it would be easy for me to learn Russian, since I already spoke Ukrainian."

As a small girl her favorite book and movie was "Heidi", because she thought it was named after her.

When Heidi was a girl, she remembers, "I really wanted to fly, but the only women that I saw on airplanes were flight attendants and I didn't see that as a very challenging career. I also wanted to be a doctor, because everyone had told me that you had to be really smart to be a doctor, and I had good grades in school. As I got into high school, I liked math and physical sciences more than biology, so I started looking at engineering."

In elementary school, Heidi was on the basketball and softball teams. She went traveling and camping with a Ukrainian Scouting organization called "Plast". She also enjoyed playing football, basketball, hockey and baseball with her brothers in their backyard. *She says, "By being active in sports, I stayed physically fit, which is important to being an astronaut. Being a Plast scout taught me teamwork and leadership."*

Heidi is married to Glenn A. Piper. They have one child, Michael, who is 12.

Her husband encouraged her to become an astronaut. "It wasn't until I had graduated from college and was in the Navy that I was serious about becoming an astronaut. I was already in my late 20's. In the early 1990's, NASA was planning on building a space station in orbit and there was talk about how many spacewalks would be required to build the station. At the time I was doing underwater ship repair, and I thought that building a space station was more like fixing ships underwater. So I decided that this was my opportunity to fly (which I had always wanted to do), so I applied to NASA."

She enjoys scuba diving, swimming, running, roller blading, and ice-skating. As an undergraduate, she competed in intercollegiate athletics on MIT's crew team.

Education:

Heidi Graduated from Derham Hall High School, St. Paul, Minnesota, in 1980. She received a Bachelor of Science degree in mechanical engineering from Massachusetts Institute of Technology in 1984, and a Master of Science degree in mechanical engineering from Massachusetts Institute of Technology in 1985.

Experience:

Stefanyshyn-Piper received her commission from the Navy ROTC Program at MIT in June 1985. She completed training at the Naval Diving and Salvage Training Center in Panama City, Florida as a Navy Basic Diving Officer and Salvage Officer.

She completed several tours of duty as an Engineering Duty Officer in the area of ship maintenance and repair and is qualified as a Surface Warfare Officer onboard U.S.S. *Grapple* (ARS 53).

In September 1994, Stefanyshyn-Piper reported to the Naval Sea Systems Command as Underwater Ship Husbandry Operations Officer for the Supervisor of Salvage and Diving. In that capacity, she advised fleet diving activities in the repair of naval vessels while waterborne.

Additionally she is a qualified and experienced salvage officer. Major salvage projects include: development of salvage plan for the Peruvian Navy salvage of the Peruvian submarine *Pacocha*; and de-stranding of the tanker *Exxon Houston*, off the coast of Barber's Point, on the island of Oahu, Hawaii.

Space Flight Experience:

Selected as an astronaut candidate by NASA in April 1996, Stefanyshyn-Piper reported to the Johnson Space Center in August 1996. Having completed two years of training and evaluation, she is qualified for flight assignment as a mission specialist. Initially assigned to astronaut support duties for launch and landing, she has also served as lead Astronaut Office Representative for Payloads and in the Astronaut Office EVA branch.

Stefanyshyn-Piper is assigned to STS-115, scheduled to launch in 2003.

Organizations:

Heidi Stefanyshyn-Piper is a member of the American Society of Mechanical Engineers.

Special Honors:

Heidi Stefanyshyn-Piper has received the "VADM C.R. Bryan Award" Class 2-88B, Engineering Duty Officer Basic Course. Awarded Meritorious Service Medal, 2 Navy Commendation Medals, 2 Navy Achievement Medals, and other service medals. "Most Valuable Player Award" MIT Women's Crew in 1982.

Astronaut Biography of Yvonne Darlene Cagle

Personal Information:
Born in West Point, New York, but considers Novato, California, to be her hometown

Yvonne's father is a retired Air Force Radiological technician and her mother is a retired Flight Operations and budget Analyst. She has two brothers and three sisters.

Yvonne first became interested in space when she was nine and watched Neil Armstrong and Buzz Aldrin walk on the moon. But it was the Tuskegee airmen and the African-American Astronauts that inspired her to become an astronaut. *As a girl she wasn't sure if she wanted to be a doctor, astronaut, or firefighter. She ended up being the first two.*

In school she enjoyed physiology and after school she was a Girl Scout and a Red Cross volunteer. She also enjoyed music and reading. Her favorite books were *The Yearling*, *Poetry of the Negro*, and literary compositions by Maya Angelou and Ida B. Wells.

As an astronaut, Yvonne wants to do, "Space and life research in order to improve our health on Earth and travel into deep space." After her first flight she wants to "Fly more and continue research into flight safety and technology, and heal whenever and wherever I encounter hurt."

Yvonne enjoys jigsaw puzzles, juggling, skating, hiking, music, writing, public speaking, and historical novels.

Education:
Yvonne Cagle graduated Novato High School Novato, California, in 1977. She received a Bachelor of Arts degree in biochemistry from San Francisco State University in 1981, and a doctorate in medicine from the University of Washington in 1985. She completed a transitional internship at Highland General Hospital, Oakland, California, in 1985. She received certification in Aerospace Medicine from the School of Aerospace Medicine at Brooks Air Force Base, Texas, in 1988. In 1992 she completed her residency in family practice at Ghent FP at Eastern Virginia Medical School. She received certification as a senior aviation medical examiner from the Federal Aviation Administration in 1995.

Experience:
Dr. Cagle's medical training was sponsored by the Health Professions Scholarship Program, through which she received her commission as an officer with the United States Air Force, and subsequently was awarded her board certification in family practice. During her initial active duty tour at Royal Air Force Lakenheath, United Kingdom, she was selected to attend the School of Aerospace Medicine at Brooks Air Force Base, Texas.

In April 1988, she became certified as a flight surgeon logging numerous hours in a variety of aircraft. She was actively involved in mission support of aircraft providing medical support and rescue in a variety of aeromedical missions.

From 1994 to 1996, Dr. Cagle served as the Deputy Project Manager for Kelsey-Seybold Clinics, practicing as an occupational physician at the NASA-JSC Occupational Health Clinic. In addition to conducting job-related exams, routine health screenings, and providing acute care for on-site injuries and illness; she designed medical protocols and conducted the screenings for select NASA remote duty operations.

Astronaut Experience:
During May 1989, while a flight surgeon assigned to the 48th Tactical Hospital, United Kingdom, Dr. Cagle volunteered to serve as the Air Force Medical Liaison Officer for the STS-30 *Atlantis* Shuttle Mission to test the Magellan Spacecraft. She was assigned to the Trans Atlantic (TAL) landing site at Banjul, West Africa, to provide emergency rescue and evacuation of the shuttle crew should it have been required.

Dr. Cagle has contributed on-going data to the Longitudinal Study on Astronaut Health, and served as a consultant for space telemedicine. She was a member of the NASA Working Group and traveled to Russia to establish international medical standards and procedures for astronauts. She also conducted health screenings of Mir-18 consultants from the Russian Federation.

Selected by NASA in April 1996, Dr. Cagle reported to the Johnson Space Center in August 1996. Having completed two years of training and evaluation, she is qualified for flight assignment as a mission specialist. Currently, Dr. Cagle is assigned technical duties in the Astronaut Office Operations Planning Branch, supporting Shuttle and Space Station. She is awaiting assignment to her first space shuttle mission.

Organizations:

Dr. Cagle is a member of the Boys and Girls Club, Aerospace Medical Association, Third Baptist Church, and American Academy of Family Physicians.

Special Honors:

Dr. Cagle has received the National Defense Service Medal, Air Force Achievement Medal, United States Air Force (USAF) Air Staff Exceptional Physician Commendation, and the National Technical Association Distinguished Scientist Award.

Astronaut Biography of Tracy E. Caldwell

Personal Information:

Born 14 August 1969 in Arcadia, California.

During high school and Undergraduate College Tracy worked as an electrician/inside wireman for her father's electrical contracting company doing commercial and light industrial type construction.

Tracy's recreational interests include running, weight training, hiking, softball, basketball, and auto repair/maintenance. She also enjoys spending time with family.

As an undergraduate, she competed in intercollegiate athletics on Cal State University Fullerton's track team as both a sprinter and long jumper.

Tracy is a private pilot and conversational in American Sign Language (ASL) and Russian.

Education:

Tracy Caldwell graduated from Beaumont High School, Beaumont, California in 1987. She received a Bachelor of Science degree in chemistry from the California State University at Fullerton in 1993 and a doctorate in physical chemistry from the University of California at Davis in 1997.

Experience:

As an undergraduate researcher at the California State University, Fullerton, Dr. Caldwell designed, constructed and implemented electronics and hardware associated with a laser-ionization, time-of-flight mass spectrometer for studying atmospherically relevant gas-phase chemistry. She also worked for the Research and Instructional Safety Office as a lab assistant performing environmental monitoring of laboratories using hazardous chemicals and radioactive materials, as well as calibrating survey instruments and helping to process chemical and radioactive waste. At the University of California, Davis, Dr. Caldwell taught general chemistry laboratory and began her graduate research. Her dissertation work focused on investigating molecular-level surface reactivity and kinetics of metal surfaces using electron spectroscopy, laser desorption, and Fourier transform mass spectrometry techniques. She also designed and built peripheral components for a variable temperature, ultra-high vacuum scanning tunneling microscopy system. In 1997, Dr. Caldwell received the Camille and Henry Drefus Postdoctoral Fellowship in Environmental Science to study atmospheric chemistry at the University of California, Irvine. There she investigated reactivity and kinetics of atmospherically relevant systems using atmospheric pressure ionization mass spectrometry, Fourier transform infrared and ultraviolet absorption spectroscopies. In addition, she developed methods of chemical ionization for spectral interpretation of trace compounds.

Astronaut Experience:

NASA selected Dr. Caldwell in June 1998. After a year of Astronaut Candidate Training she was qualified as an astronaut. Dr. Caldwell is currently assigned to the Astronaut Office Space Station Operations Branch. She will serve in technical assignments until assigned to a space flight.

Organizations:

Dr. Caldwell is a member of the Sigma Xi Research Society and American Association for the Advancement of Science.

Astronaut Biography of Joan E. Higginbotham

Personal Information:

Born 3 August in Chicago, Illinois.

Joan grew up in Chicago with her four siblings and working parents.

"In high school, I knew I wanted to double up on my math requirements in my sophomore year so I could begin calculus before college," Joan told a crowd at the January 2001 Building Careers for Maui Conference. "When my guidance counselor pointed out the statistically negative success rate of that choice, it didn't sway me. My mother approved my choice and I did well in both classes. *I have never felt like a 'statistic,' knowing I could handle whatever I put my mind to. When someone says 'no' to me, I work all the harder to prove them wrong.*"

Joan applied but was not selected in the astronaut class of 1994. Her perseverance paid off, however, when she applied again and was selected in 1996.

Joan enjoys weightlifting, cycling, music, and motivational speaking.

Education:

Joan Higginbotham graduated from Whitney M. Young Magnet High School, Chicago, Illinois, in 1982. She received a Bachelor of Science degree electrical engineering from Southern Illinois University at Carbondale, in 1987, and masters of management and space systems from Florida Institute of Technology in 1992 and 1996, respectively.

Experience:

Joan Higginbotham began her career in 1987 at the Kennedy Space Center (KSC), Florida, as a Payload Electrical Engineer in the Electrical and Telecommunications Systems Division. Within six months she became the lead for the Orbiter Experiments (OEX) on OV-102, the Space Shuttle *Columbia*. She later worked on the Shuttle payload bay reconfiguration for all Shuttle missions and conducted electrical compatibility tests for all payloads flown aboard the Shuttle. She was also tasked by KSC management to undertake several special assignments where she served as the Executive Staff Assistant to the Director of Shuttle Operations and Management, led a team of engineers in performing critical analysis for the Space Shuttle flow in support of a simulation model tool, and worked on an interactive display detailing the Space Shuttle processing procedures at Spaceport USA (Kennedy Space Center's Visitors Center). Higginbotham then served as backup orbiter project engineer for OV-104, Space Shuttle *Atlantis*, where she participated in the integration of the orbiter docking station (ODS) into the space shuttle used during Shuttle/Mir docking missions. Two years later, she was promoted to lead orbiter project engineer for OV-102, Space Shuttle *Columbia*. In this position, she held the technical lead government engineering position in the firing room where she supported and managed the integration of vehicle testing and troubleshooting. She actively participated in 53 space shuttle launches during her nine-year tenure at the Kennedy Space Center.

Astronaut Experience:

Selected as an astronaut candidate by NASA in April 1996 Joan Higginbotham reported to the Johnson Space Center in August 1996. Since that time, she has been assigned technical duties in the Payloads & Habitability Branch, the Shuttle Avionics & Integration Laboratory (SAIL), and the Kennedy Space Center (KSC) Operations (Ops) Support Branch. In her previous assignment with the KSC Ops Support Branch, she tested various modules of the International Space Station for operability, compatibility, and functionality prior to launch.

Currently, Joan Higginbotham is assigned technical duties in the Astronaut Office CAPCOM (Capsule Communicator) Branch. Having completed two years of training and evaluation, she is qualified for flight assignment as a mission specialist.

Organizations:

Higginbotham is a member of the Delta Sigma Theta Sorority, Inc., Bronze Eagles, Links Inc.

Special Honors:

Her awards include the NASA Exceptional Service Medal, Keys to the Cities of Cocoa and Rockledge, Florida; Group Achievement Award for STS-26 Return to Flight, Kennedy Space Center Public Affairs Certificate of Appreciation for Service and a Commendation of Merit for Service to the Department of Defense (DOD) Missions.

Astronaut Biography of Nadezhda Kuzhelnaya

Personal Information:

Born 11 June 1962 in Alexeevskaya, a town located in Volga region, in central Russia.

"I wanted to become a cosmonaut like many other Russian kids, having read a lot of science fiction literature and having seen many science fiction movies," Nadezhda told Space.com. "I was also very excited by Savitskaya's flight."

Nadezhda has a passion for flying. In 1981, she joined an aviation club and flew aerobatic aircraft. She became a member of the *Swallows*, an all-female Russian aerobatics team.

"I would not say that female cosmonauts are not welcomed in the Russian space program," says Kuzhelnaya. "I must say however, that all spaceflight hardware, including spacesuits and spacecraft comfort assuring systems, were designed mostly by men and for men. For this reason women do not really fit into the Russian spacecraft environment."

Nadezhda is married to Vladimir Kuzhelnaya, who was her flight instructor at Star City. They have one young daughter, Katya. Nadezhda's hobbies are playing acoustic guitar and singing.

Education:

Nadezhda Kuzhelnaya graduated from high school in Krivoi Rog, Ukraine. She studied engineering from 1981 to 1984 at the Dnepropetrovsk engineering Institute and graduated from the Moscow Aviation Institute in 1988.

Experience:

After graduation she was an engineer at Energia, Russia's leading manufacturer of space hardware.

Rosaviakosmos (The Russian Aviation and Space Agency) selected Kuzhelnaya for Cosmonaut training on 4 January 1994.

Kuzhelnaya was scheduled to fly on a *Soyuz* from Baikonur in late April 2001, but by the end of 2000 she was pulled from the schedule. The official reason for Kuzhelnaya's removal from this mission was her height. She is reportedly not large enough to work in the Russian EVA suit. This is the same reason Wendy Lawrence was pulled from her mission to *Mir* several years ago. Kuzhelnaya, however, was able to fit into the EVA suit and perform tasks in the Russian neutral buoyancy tank.

"I did all the required procedures during my training in the neutral buoyancy tank. Besides, none of the 'taxi' mission crew members is expected to do a spacewalk," she told Space.com. "Unfortunately, if a cosmonaut does not fit into a spacesuit, it is easier for the Russian space authorities to find a replacement for him or for her, than for a spacesuit. *I don't think that until such attitude is changed, there will be more women cosmonauts in Russia. Unfortunately, this tendency reflects an old tradition in Russian society where women usually are subordinate to men."*

Kuzhelnaya is awaiting assignment for another mission. "I know that American EVA suits are more comfortable than the Russians." I have a slight hope that maybe some day I will work outside a spacecraft in a U.S. EVA spacesuit. I would be very happy to fly a main mission in ISS, or as a mission, or payload specialist on the space shuttle."

Astronaut Biography of K. Megan McArthur

Personal Information:

Born on 30 August 1971 in Honolulu, Hawaii, but considers California to be her home state.

While at school, Megan spent her free time with educational outreach projects and community service projects. She worked with the University of California at San Diego's Biology Department programs aimed at teaching ethnically diverse school children about the excitement of aerospace and ocean engineering.

She volunteered to help out with the National Science Foundation's "Girl Power" program, a summer enrichment program for young girls that does experiments to show the fun side of math and science.

In addition, Megan also volunteered at the Birch Aquarium at Scripps, conducting educational demonstrations for the public from inside a 70,000-gallon exhibit tank of the California Kelp Forest. The demonstrations included in-water tank maintenance, feeding and observation of the animals, and Q&A with the audience. Megan has also been a volunteer for Scripps CHiPS (Committee for Humanity and Public Service) a group that does environmental clean up and community work.

Megan is the youngest of the active astronauts. She told Space.com, "Sometimes they call me the baby, but I don't mind." However, she's not the youngest astronaut selected by NASA. That was Tammy Jernigan (age 26 when selected). And of course Valentina Tereshkova was only 25 when she began her cosmonaut training.

Megan enjoys scuba diving, backpacking, and cooking. Her parents, Don & Kit McArthur, reside in San Jose, California. Her brother, Sean McArthur, and sisters Shannon McArthur Silva and Erin Dahlby, also reside in California.

Education:

Megan graduated from St. Francis High School, Mountain View, CA, 1989. Bachelor of Science in Aerospace Engineering, University of California at Los Angeles in 1993 and a Ph.D. in Oceanography, University of California in San Diego, 2001.

Experience:

At the Scripps Institution of Oceanography, Megan conducted graduate research in near-shore underwater acoustic propagation and digital signal processing. Her research focused on determining geo-acoustic models to describe very shallow water wave-guides using measured transmission loss data in a genetic algorithm inversion technique. She served as Chief Scientist during at-sea data collection operations, and has planned and led diving operations during sea-floor instrument deployments and sediment-sample collections. While at Scripps, she participated in a range of in-water instrument testing, deployment, maintenance, recovery, and collection of marine plants, animals, and sediment.

Astronaut Experience:

Selected as a Mission Specialist candidate by NASA in July 2000, Megan McArthur reported for astronaut basic training in August 2000. She will serve in technical assignments until assigned to a space flight.

Astronaut Biography of Barbara R. Morgan

Personal Information:

Born 28 November 1951, in Fresno, California.

Barbara is married to writer Clay Morgan of McCall, Idaho. They have two sons. She is a classical flautist who also enjoys jazz, literature, hiking, swimming, cross-country skiing, and spending time with her family.

Barbara was the backup Teacher in Space for Christa McAuliffe. After the 1986 *Challenger* disaster, she traveled around the U.S. for three months speaking with schoolchildren about the astronauts who died and the future of space exploration. Barbara returned to her teaching job in Idaho and was an educational consultant for NASA, traveling around the U.S. speaking for the space agency.

After years of applying and re-applying to the Astronaut Office, Barbara was finally chosen to be an astronaut candidate in 1998. An example of someone who persevered and followed her dreams, Barbara will finally get her chance to fly in space in 2004 Her flight will be the first in NASA's new Educator Mission Specialist Program. The space agency plans to work with the Department of Education to recruit more teachers to fly on the space shuttle.

Education:

Barbara Morgan graduated Hoover High School, Fresno, California in 1969. She received a Bachelor of Arts in Human Biology from Stanford University in 1973 (with distinction), and her teaching Credential from College of Notre Dame, Belmont, California in 1974.

Experience:

Morgan began her teaching career in 1974 on the Flathead Indian Reservation at Arlee Elementary School in Arlee, Montana, where she taught remedial reading and math. From 1975-1978, she taught remedial reading/math and second grade at McCall-Donnelly Elementary School in McCall, Idaho. From 1978-1979, Morgan taught English and science to third graders at Colegio Americano de Quito in Quito, Ecuador. From 1979-1998, she taught second, third, and fourth grades at McCall-Donnelly Elementary School.

Astronaut Experience:

Morgan was selected as the backup candidate for the NASA Teacher in Space Program on July 19, 1985. From September 1985 to January 1986, Morgan trained with Christa McAuliffe and the *Challenger* crew at NASA's Johnson Space Center, Houston, Texas. Following the *Challenger* accident, Morgan assumed the duties of Teacher in Space

Designee. From March 1986 to July 1986, she worked with NASA, speaking to educational organizations throughout the country. In the fall of 1986, Morgan returned to Idaho to resume her teaching career. She taught second and third grades at McCall-Donnelly Elementary and continued to work with NASA's Education Division, Office of Human Resources and Education. Her duties as Teacher in Space Designee included public speaking, educational consulting, curriculum design, and serving on the National Science Foundation's Federal Task Force for Women and Minorities in Science and Engineering. Selected by NASA in January 1998 as the first Educator Mission Specialist, she reported for training in August 1998. Morgan is currently assigned to the Astronaut Office Space Station Operations Branch. She will serve in technical assignments until assigned to a space flight. NASA administrator O'Keefe recently announced her first flight will be in 2004.

Organizations:

Morgan is a member of the Idaho Education Association, National Council of Teachers of Mathematics, National Science Teachers Association, International Reading Association, International Technology Education Association, and *Challenger* Center for Space Science Education.

Special Honors:

Morgan's awards include Phi Beta Kappa (1973), NASA Headquarters Special Service Award (1987), NASA Public Service Group Achievement Award (1988). Other awards include Idaho Fellowship Award (1998), University of Idaho President's Medallion Award (1998), and the Women in Aerospace Education Award (1991).

Astronaut Biography of Lisa M. Nowak

Personal Information:

Born 10 May 1963, in Washington, DC.

Lisa is married to Richard T. Nowak of South Burlington, Vermont. They have three children. She enjoys bicycling, running, skeet, sailing, gourmet cooking, rubber stamps, crossword puzzles and piano. As an undergraduate, Lisa competed on the track team.

Education:

Lisa Nowak graduated from C.W. Woodward High School, Rockville, Maryland, in 1981. She received a Bachelor of Science degree in aerospace engineering from the U.S. Naval Academy in 1985. She received a Master of Science degree in aeronautical engineering and a degree of aeronautical and astronautical engineer from the U.S. Naval Postgraduate School, both in 1992.

Experience:

Nowak received her commission from the U.S. Naval Academy in May 1985, and reported to flight school after six months of temporary duty at Johnson Space Center. She earned her wings as a Naval Flight Officer in June 1987, followed by Electronic Warfare School at Corry Station, Florida, and initial A-7 training at Naval Air Station Lemoore, California.

She was assigned to Electronic Warfare Aggressor Squadron 34 at Point Mugu, California, where she flew EA-7L and ERA-3B aircraft, supporting the fleet in small and large-scale exercises with jamming and missile profiles. While assigned to the squadron, she qualified as Mission Commander and EW Lead. In 1992, Nowak completed two years of graduate studies at Monterey, and began working at the Systems Engineering Test Directorate at Patuxent River, Maryland. In 1993, she was selected for both Aerospace Engineering Duty and U.S. Naval Test Pilot School. After graduation in June 1994, she stayed at Patuxent River working as an aircraft systems project officer at the Air Combat Environment Test and Evaluation Facility and at Strike Aircraft Test Squadron, flying the F/A-18 and EA-6B. Nowak was then assigned to the Naval Air Systems Command, working on acquisition of new systems for naval aircraft, when she was selected for the astronaut program.

Nowak has logged over 1,100 flight hours in more than 30 different aircraft.

Astronaut Experience:

After receiving her commission, Commander Nowak was assigned temporary duty and from June to November 1985 she provided engineering support for JSC's Shuttle Training Aircraft Branch at Ellington, Texas. Selected as a NASA astronaut candidate in April 1996, Nowak reported to the Johnson Space Center in August 1996. Having completed two years of training and evaluation, she is qualified for flight assignment as a mission specialist. Initially assigned technical duties in the Astronaut Office Operations Planning Branch, Nowak is currently assigned to the Astronaut Office CAPCOM Branch, working in Mission Control as prime communicator with on-orbit crews.

Organizations:

Commander Nowak is a member of the American Institute of Aeronautics and Astronautics (AIAA). U.S. Naval Academy Alumni Association, and Tau Beta Pi Engineering Society.

Special Honors:

Commander Nowak has received the Navy Commendation Medal. Navy Achievement Medal, and various other service awards.

Astronaut Biography of Karen L. Nyberg

Personal Information:

Born 7 October 1969 in Parkers Prairie, Minnesota. Her hometown is Vining, Minnesota.

Her dad, Ken Nyberg of Vining Minnesota, creates painted steel sculptures — from scrap metal. One is a sculpture of his astronaut daughter.

Karen's recreational interests include art, running, volleyball, sewing, backpacking, piano, and spending time with her dogs.

Education:

Karen Nyberg graduated from Henning Public High School, Henning, Minnesota, 1988. She received a Batchelor of Science of Mechanical Engineering, University of North Dakota in 1994. She received a Master of Science in Mechanical Engineering, University of Texas at Austin in 1996 and a Ph.D. in Mechanical Engineering, University of Texas at Austin in 1998.

Experience:

Dr. Nyberg completed graduate research at The University of Texas at Austin in the BioHeat Transfer Laboratory where she investigated human thermoregulation and experimental metabolic testing and control, specifically related to the control of thermal neutrality in space suits.

Dr. Nyberg was a co-op at Johnson Space Center from 1991-1995, working in a variety of areas. She received a patent for work done in 1991 on Robot Friendly Probe and Socket Assembly.

In 1998, on completing her doctorate, she accepted a position with the Crew and Thermal Systems Division, working as an Environmental Control Systems Engineer. Her prime responsibility involved using human thermal physiology and engineering control for improvements in the space suit thermal control system and evaluation of firefighter suit cooling technologies. Other responsibilities included providing computational fluid dynamic analysis for the TransHab module air distribution system, coordinating and monitoring analysis tasks performed by a team of contractor personnel for the X-38 environmental control and life support systems, providing conceptual designs of the thermal control system for the Advanced Mars and Lunar Lander Mission studies, and environmental control system analysis for a collapsible hyperbaric chamber.

Astronaut Experience:

Selected as a mission specialist by NASA in July 2000, Dr. Nyberg reported for training in August 2000. Following initial training, she will serve in technical assignments until assigned to a space flight.

Organizations:

Dr. Nyberg is a member of the Tau Beta Pi (National Engineering Honor Society), American Society of Mechanical Engineers, and Aerospace Medical Association.

Special Honors:

Her awards include the Space Act Award (1993), NASA Tech Briefs Award (1993), NASA JSC Cooperative Education Special Achievement Award (1994), and the Joyce Medalen Society of Women Engineers Award (1993-94).

Astronaut Biography of Nicole Passonno Stott

Personal Information:

Born 19 November 1962 in Albany, New York. But considers Clearwater, Florida her hometown.

As a teen, Nicole dreamed of flying. She had a passion for flying and helped her father build biplanes in the garage.

Tragedy struck Nicole's life when she was almost 16. Her father died when an experimental plane he was flying crashed. "I was devastated when it happened of course, but I never thought about giving up flying," she told the St. Petersburg Times. "He was a guy whose whole life was spent flying and building planes, and the way he died was kind of the culmination of that."

The crash didn't end her interest in flying. She got her pilot's license at age 18 and pursued aviation on to the highest form of flying; she became an astronaut. As she told the paper, "I never pictured myself as an astronaut when I was growing up. It wasn't my lifelong dream or anything. But then again, I'd never not pictured myself doing it."

"She's just one of these people you like to work with," said Tip Talone, director of International Space Station and payload processing at Kennedy Space Center. "She's mature and outgoing, but also a team player and she knows the program inside and out. When you add those qualities together, she fits like a hand in a glove."

Nicole said, "It took me a long time to figure out I could make career out of doing something I love."

Nicole is married to Christopher Stott of the Isle of Man, UK. They live in Houston with their two dogs, Stella and Lucy. Nicole enjoys flying, snow skiing, scuba diving, woodworking, painting, and gardening.

Education:

She graduated Clearwater High School, Clearwater, Florida in 1980 and received her Bachelor of Science in Aeronautical Engineering from the Embry-Riddle Aeronautical University in 1987. She received her Masters of Science in Engineering Management, University of Central Florida in 1992.

Experience:

Ms. Stott began her career in 1987 as a structural design engineer with Pratt and Whitney Government Engines in West Palm Beach, FL. She spent a year with the Advanced Engines Group performing structural analyses of advanced jet engine component designs.

In 1988, Ms. Stott joined NASA at the Kennedy Space Center (KSC), Florida as an Operations Engineer in the Orbiter Processing Facility (OPF). After 6 months, she was detailed to the Director of Shuttle Processing as part of a two-person team tasked with assessing the overall efficiency of Shuttle processing flows, identifying and implementing process improvements, and implementing tools for measuring the effectiveness of improvements. She was the NASA KSC Lead for a joint Ames/KSC software project to develop intelligent scheduling tools. The Ground Processing Scheduling System (GPSS) was developed as the technology demonstrator for this project. GPSS was a success at KSC and is still in use today for OPF scheduling, and also a commercial success that is part of the PeopleSoft suite of software products. During her time at KSC, Ms. Stott also held a variety of positions within NASA Shuttle Processing, including Vehicle Operations Engineer; NASA Convoy Commander; Shuttle Flow Director for Endeavour; and Orbiter Project Engineer for *Columbia*.

During her last two years at KSC, she was a member of the Space Station Hardware Integration Office and relocated to Huntington Beach, CA where she served as the NASA Project Lead for the ISS truss elements under construction at the Boeing Space Station facility. In 1998, she joined the Johnson Space Center (JSC) team in Houston, TX as a member of the NASA Aircraft Operations Division, where she served as a Flight Simulation Engineer (FSE) on the Shuttle Training Aircraft (STA).

Astronaut Experience:

NASA selected Stott as a mission specialist in July 2000, Nicole Stott reported for astronaut candidate training in August 2000. Following initial training, she will serve in technical assignments until assigned to a space flight.

Organizations:

Stott is a member of the Aircraft Owners and Pilot's Association (AOPA), and The Mars Society.

Special Honors:

She has received awards including the Newt Myers Team Spirit Award (2000), KSC Public Affairs Certificate of Appreciation for Service (1998), NASA Exceptional Achievement Medal (1997), three NASA Certificates of Commendation (1993, 1995, 1998), and nine NASA Performance Awards (1990, 1992, 1993, 1995, 1996, 1997, 1998, 1999, 2000).

Astronaut Biography of Naoko Sumino

Personal Information:

Born 27 December 1970 in Matsudo, Chiba, Japan

Naoko and her husband are expecting their first baby in August 2002. "I would like to strike a balance between my work and child-rearing after I give birth to the baby," she said at a news conference at the Tsukuba Space Center in Ibaraki, Japan. "I believe the current training for space flight has been a good prenatal influence on the baby."

Her training schedule is being adjusted to account for the pregnancy. After maternity leave Naoko plans to return to work. There are several female astronauts with children in the United States, and she does not anticipate future problems in preparing or participating in space flight.

Education:

Sumino graduated from an all-girl high school, which is attached to Ochanomizu University. She received her Bachelor of Science in Aerospace Engineering from the University of Tokyo in 1993, and her Master of Science in Aerospace Engineering also from the University of Tokyo in 1996.

Experience:

Following graduation Sumino was an engineer on the Japanese Experiment Module Project Team, designing the Japanese module for the International Space Station (ISS) from April 1996 to May 1998. From June 1998 to March 1999, Sumino was an engineer on the National Space Development Agency (NASDA - Japan's space agency) centrifuge project team.

Astronaut Experience:

In April 1999, she was selected by NASDA as an astronaut and enrolled in the NASDA basic training program for Japanese astronauts. She is the second Japanese woman astronaut. She completed examinations and training and was qualified as an NASDA ISS astronaut on 26 September 2001. Sumino is part of the first group of Japanese astronauts trained in Japan rather than by NASA in the United States.

Sumino started advanced ISS training after the basic training. The advanced training takes place in several countries. She is also supporting the development of the Japanese Experiment Module (Kibo). Sumino is awaiting assignment to an ISS flight.

Astronaut Biography of Sunita L. Williams

Personal Information:

Born 19 September 1965 in Euclid, Ohio, but considers Needham, Massachusetts to be her hometown.

Married to Michael J. Williams. They have no children, but they have an "awesome" black Labrador retriever named Turbo.

Her recreational interests include running, swimming, biking, triathlons, windsurfing, snowboarding and bow hunting.

Education:

Sunita Williams graduated from Needham High School, Needham, Massachusetts in 1983. She received her Bachelor of Science in Physical Science from the U.S. Naval Academy in 1987. She received her Master of Science in Engineering Management from the Florida Institute of Technology in 1995.

Experience:

Williams received her commission as an Ensign in the United States Navy from the U.S. Naval Academy in May 1987. After a six-month temporary assignment at the Naval Coastal System Command, she received her designation as a Basic Diving Officer and then reported to Naval Aviation Training Command. She was designated a Naval Aviator in July 1989.

She then reported to Helicopter Combat Support Squadron 3 for initial H46, Sea-Knight, training. Upon completion of this training, she was assigned to Helicopter Combat Support Squadron 8 in Norfolk, Virginia, and made overseas deployments to the Mediterranean, Red Sea and the Persian Gulf in support of Desert Shield and Operation Provide Comfort. In September 1992 she was the Officer-in-Charge of an H-46 detachment sent to Miami, Florida for Hurricane Andrew Relief Operations onboard USS Sylvania.

Williams was selected for United States Naval Test Pilot School and began the course in January 1993. After graduation in December 1993, she was assigned to the Rotary Wing Aircraft Test Directorate as an H-46 Project Officer, and V-22 Chase Pilot in the T-2. While there she was also assigned as the squadron Safety Officer and flew test flights in the SH-60B/F, UH-1, AH-1W, SH-2, VH-3, H-46, CH-53 and the H-57. In December 1995, she went back to the Naval Test Pilot School as an Instructor in the Rotary Wing Department and the school's Safety Officer. There she flew the UH-60, OH-6 and the OH-58. From there she was assigned to the USS Saipan (LHA-2), Norfolk, Virginia, as the Aircraft Handler and the Assistant Air Boss. Williams was deployed onboard USS Saipan when she was selected for the astronaut program. She has logged over 2300 flight hours in more than 30 different aircraft.

Astronaut Experience:

NASA selected Lieutenant Commander Williams in June 1998. Williams is currently assigned to the Astronaut Office Space Station Operations Branch where she will serve in technical assignments until assigned to a space flight.

Organizations:

Lieutenant Commander Williams is a member of the Society of Experimental Test Pilots, Society of Flight Test Engineers, and American Helicopter Association.

Special Honors:

Lieutenant Commander Williams was awarded the Navy Commendation Medal twice, Navy and Marine Corps Achievement Medal, Humanitarian Service Medal and various other service awards.

Astronaut Biography of Stephanie D. Wilson

Personal Information:

Born in 1966 in Boston Massachusetts.

"This is a dream come true for me," Stephanie said after her selection. "I am very excited and honored to have been selected from so many qualified candidates. I believe the experience I have gained here at JPL, working with a spacecraft exploring the solar system, has best prepared me for the astronaut program."

Stephanie is married to Julius "BJ" McCurdy. She enjoys snow skiing, music, astronomy, stamp collecting, and traveling.

Education:

Stephanie Wilson graduated from Taconic High School, Pittsfield, Massachusetts, in 1984. She received a Bachelor of Science degree in engineering science from Harvard University in 1988, and a Master of Science degree in aerospace engineering from the University of Texas, in 1992.

Experience:

After graduating from Harvard in 1988, Wilson worked for 2 years for the former Martin Marietta Astronautics Group in Denver, Colorado. As a Loads and Dynamics engineer for Titan IV, Wilson was responsible for performing coupled load analyses for the launch vehicle and payloads during flight events.

Wilson left Martin Marietta in 1990 to attend graduate school at the University of Texas. Her research focused on the control and modeling of large, flexible space structures. Following the completion of her graduate work, in 1992 she began working for the Jet Propulsion Laboratory in Pasadena, California. As a member of the Attitude and Articulation Control Subsystem for the Galileo spacecraft,

Wilson was responsible for assessing attitude controller performance, science platform pointing accuracy, antenna pointing accuracy and spin rate accuracy. She worked in the areas of sequence development and testing as well. While at the Jet Propulsion Laboratory, Wilson also supported the Interferometery Technology Program as a member of the Integrated Modeling Team, which was responsible for finite element modeling, controller design, and software development.

Astronaut Experience:

NASA selected Wilson in April 1996 and she started training at the Johnson Space Center that August. Having completed two years of training and evaluation, she is qualified for flight assignment as a mission specialist. Initially assigned technical duties in the Astronaut Office Space Station Operations Branch, Wilson is currently assigned to the Astronaut Office CAPCOM Branch, working in Mission Control as prime communicator with on-orbit crews.

Organizations:

Wilson is a member of the American Institute of Aeronautics and Astronautics (AIAA).

Astronaut Biography of Patricia Hilliard Robertson

Personal Information:

Born 12 March 1963 in Indiana, Pennsylvania.

Patricia died May 24, 2001 of injuries sustained in the crash of a single-engine experimental private plane at Wolfe Air Park, Manvel, Texas. Witnesses said the plane, carrying Robertson and pilot and aircraft owner Roy Mack Paul Adams, cartwheeled after practicing landing and takeoff maneuvers.

She was married to Scott Robertson.

A tree is planted in Patty's name, as there is for each fallen astronaut, at NASA's Johnson Space Center in Houston.

Patty was a multi-engine rated flight instructor and avid aerobatic pilot. In her free time, she enjoyed flight instructing, aerobatics, and flying with her husband. She had accumulated over 1500 hours of flight time.

Education:

Patricia Robertson graduated from Homer Center High School, Homer City, Pennsylvania In 1980. She received a Bachelor of Science degree in biology from Indiana University of Pennsylvania in 1985, and a medical degree from the Medical College of Pennsylvania in 1989. She completed a three-year residency in Family Medicine in 1992 and was certified by the American Board of Family Practice in the same year. In 1997 she completed a two-year fellowship in Space Medicine at the University of Texas Medical Branch and NASA Johnson Space Center, which included the Aerospace Medicine Primary Course at Brooks Air Force Base.

Experience:

After completing her training in Family Medicine in 1992, Dr. Robertson joined a group practice in Erie, Pennsylvania. She was on the staff of St. Vincent hospital for three years where she served as the clinical coordinator for medical student training, and also provided training and supervision for resident physicians. In 1995 Dr. Robertson was one of two fellows selected to study aerospace medicine at the University of Texas Medical Branch (UTMB), Galveston, and at the Johnson Space Center, Houston. While enrolled as a space medicine fellow, Dr. Robertson completed a research project where she studied eccentric and concentric resistive exercise countermeasures for space flight. Dr. Robertson also served as a member of the faculty at UTMB in the Departments of Family Medicine and Emergency Medicine. In 1997, Dr. Robertson joined the Flight Medicine Clinic at Johnson Space Center, where she provided health care for astronauts and their families, and served as Chairman of the Bone, Muscle, and Exercise Integrated Product Team.

Astronaut Experience:

Selected by NASA in June 1998, Dr. Robertson reported for training in August 1998. Astronaut Candidate Training included orientation briefings and tours, numerous scientific and technical briefings, intensive instruction in Shuttle and International Space Station systems, physiological training and ground school to prepare for T-38 flight training, as well as learning water and wilderness survival techniques. Among her technical assignments she served as the office representative for the Crew Healthcare System (CHeCS), and as Crew Support Astronaut (CSA) for the ISS Expedition 2 Crew. Dr. Robertson never got the chance to fly into space.

Organizations:

Dr. Robertson was a member of the Aerospace Medicine Association, American Association of Family Practice, Experimental Aircraft Association, International Aerobatic Club, and Aircraft Owners and Pilot's Association.

5

Space Medicine: Women and Zero-G

As on Earth, space health issues are nearly identical between genders. But, there are differences between women and men, and some women's health issues have been studied separately. This chapter is a clinical look at some of those unique health aspects that may make living in space just a little bit different for women.

Selecting Women Astronauts

The only area where NASA makes a distinction between male and female astronaut applicants is in medicine. There is one additional medical test that women must pass to become an astronaut candidate, the pelvic exam.

Nearly identical to the annual exam that every woman should have, female astronaut candidates are tested to make sure there are no gynecological problems that would cause difficulties during her training or mission in space. In addition to the standard pelvic exam, doctors do an ultrasound of the reproductive organs to make sure there are no unusual conditions that would cause her to have medical problems while training or in space. If something unusual is found, the astronaut candidate has the option of getting treatment for her condition and taking the test again.

The Menstrual Cycle and Living in Space.

The menstrual cycle sets women apart from men. Throughout history, menstruation has often been the excuse that has kept an otherwise qualified woman from doing what she wanted to do with her life. Women weren't allowed to do certain jobs because of fears that their menstrual cycle would get in the way. For much of history, women were considered fragile. One would think childbirth, which is physically difficult in most cases, would have dispelled this myth.

Since 1960, women have had access to medications that have the effect of lessening, regulating, or eliminating menstrual flow and associated pain and discomfort. For those who see it as a problem, modern medications seem to have liberated women from the "burden" of menstruation.

Women are now able to use modern science to better their lives. They are no longer required to overcome something that is a natural part of being female just to get along in the workplace. If a woman wants to control her menstrual cycle with chemicals prescribed by her doctor, then she has that option. Some women have premenstrual syndrome, which is in some cases debilitating, but medication can help. Some women have mild cramps. Whatever the "problem," it is her choice to deal with it as she pleases.

NASA had concerns about women menstruating in space as late as 1982. It had been four years since the first U.S. women astronaut candidates were chosen, and a year before Sally Ride's first flight. Medical consultants met with NASA doctors to discuss the matter and came to the conclusion that the answer to the "problem" of menstrual flow was drugs that can reduce, delay, or stop a woman's menstrual flow.

However, to NASA's credit, the space agency does not require, and has never required, its women astronauts to "fix" their menstrual cycle. They are given the option to delay or suppress their menstrual cycle by taking medication. Some women astronauts do, and some don't.

A woman's menstrual cycle really doesn't make a difference when it comes to living in space. Some women astronauts have had their period in space and used the same sanitary products that are available here on Earth. Astronauts say they work just as well in space as they do on the ground.

Astronaut Janice Voss says that dealing with her period in space was, "very similar to on the ground. Blood doesn't flow down, so the first day was more of a surprise," says Voss. "And the urine filters in the toilet have to be changed more frequently."

Astronaut and medical doctor Ellen Baker wasn't interested in controlling her cycle either. "You just deal with it on orbit like you would deal with it on any trip. It's never interfered with anything I've done," says Baker. "I've never let it interfere with anything I've done, nor come into any decision making process that I have. So, I just decided I've lived with it all of these years, I can live with it in space."

Endometriosis

Women who choose to have their period in space may have a higher risk of developing endometriosis. Ellen Baker says that endometriosis is a common problem for women here on Earth. The lack of gravity might increase the chances, but there are so few women astronauts that it is impossible to do a proper statistical sampling. It is unclear whether zero gravity really increases the risk of endometriosis or not.

What is endometriosis? The inner lining of the uterus, called the *endometrium*, is shed during menstruation. If a woman is pregnant, the endometrium thickens with blood vessels that support the growing fetus. Endometriosis occurs when this tissue is found outside the uterus in other areas of the body such as the pelvis. It is painful and can cause internal bleeding.

On long-duration missions to Mars and beyond, women astronauts may have to take some medical precautions before leaving Earth. Preventing endometriosis will be important as people begin spending more time in low gravity and zero-g. Fortunately, there are many ways to treat, and possibly prevent the disease.

Baker says that endometriosis is not a major issue right now. "Maybe in 50 years if we're on Mars or living on the Moon there will be other issues to consider. But, right now spaceflight is relatively short duration on the shuttle or even six months on the space station. So the ill effects are fairly minor in the long term."

Radiation

Long trips into space may expose space travelers to excessive radiation. Women and men are both susceptible to cancer induced by radiation, but in different ways.

Radiation damages cells and can often lead to cancer later in life. The National Council on Radiation Protection and Measurements (NCRP) sets the maximum lifetime radiation that people should not exceed. This is the amount that will cause a 3% increase in the chance of cancer later in life. These limits are different for women because breast tissue is easily damaged by radiation.

A woman would have a higher chance of breast cancer if she were exposed to high levels of radiation. However, the chances of getting cancer from radiation depend much more on a person's age than on gender. The younger a person is when exposed to radiation, the higher the risk of cancer later in life.

Radiation limits for humans are measured in REM, which stands for "Roentgen Equivalent Man." The maximum annual dose of radiation set by the NCRP is 50 REM per year or 25 REM per month for either a man or woman at any age. The maximum lifetime radiation limit is 50 REMs for a woman at age 25, and 80 REMs for a man of the same age.

The maximum lifetime dosage goes up with age. At age 45, the dose for women is 130 REM and for men 200 REM. However, the faster a person's body absorbs the dose of radiation, the more damaging it will be.

Because the ISS is orbiting inside Earth's protective magnetic shield, the expected dosage, on even the longest missions to the International Space Station (ISS), is well below the limit established for both men and women.

The Earth from Space as seen from beyond Earth's magnetic shield

Radiation levels in space increase with solar activity. During a solar flare, radiation levels can increase many thousands of times above normal levels. Ground bases monitoring sunspot activity can see the flare, and predict the oncoming radiation bombardment before it arrives at Earth or anywhere else in the solar system. This is not so much a problem for astronauts on the Shuttle or on the ISS. It would be a problem, however, for astronauts traveling to the Moon, Mars, or beyond.

Using current technology, it may take up to a year to travel to Mars and a year to return to Earth. A typical trip to Mars would expose crewmembers to 50 REM per year. This is the maximum annual dose for both men and women. The danger is that a solar flare could result in radiation levels well beyond the maximum recommended dose. This would greatly increase an astronaut's chance of getting cancer later in life.

The problem is not that the radiation is too high for a *woman* astronaut to travel to Mars. The difficulty is that without a lot of radiation shielding on the spacecraft, radiation levels are potentially too high for *anyone* to travel to Mars and back.

Fertility and Pregnancy

Like other parts of the body, the reproductive organs are exposed to radiation. Men are actually at a higher risk of damage to the reproductive system than women. Because a woman's reproductive organs are more protected inside the body, radiation must travel through more tissue to do damage.

Fertility is not a problem for astronauts today. The short missions on the space shuttle and to the ISS do not affect an astronaut's ability to have children when they return to Earth.

Like many professional women, several astronauts have waited until their late '30's or early '40's to have their first baby. Although there is a higher chance of complications, medical technology is lessening the danger. "I think that the reproductive issues are those that anybody in a very active and busy career might face," says Baker. If a woman waits until age 40 to start a family then there are, "problems associated with that, regardless of her profession."

Flying in space while pregnant would be too risky. A fetus is very susceptible to radiation and should not be exposed to more than a total of 0.5 REM. So pregnancy in space doesn't seem like a very good idea, at least for now.

Women astronauts are tested for pregnancy 10 days before launch and again 2 days before launch. No one knows how launch, radiation and microgravity would damage a growing fetus, and childbirth in microgravity would cause countless problems. The radiation a woman absorbs while on an airplane flight in her first few weeks of pregnancy is considered potentially damaging to a fetus. So, increasing the dose, as you would in spaceflight, would be very dangerous. Because of the danger that launch, zero gravity, and radiation would pose to a growing fetus, a pregnant crewmember would have to be replaced with a backup crewmember.

In the future, when women and men live in space colonies on other worlds, gravity and other environmental issues will cause all sorts of challenges for those wanting to become parents. But right now, the effects of space travel on fertility and future pregnancies are small, and women astronauts are able to deal with them here on Earth.

Orthostatic Intolerance

When an astronaut first returns to Earth, it takes time for her body to adjust to normal gravity again. Ellen Baker says, "Really, you're not able to stand up very well the first hour that you're back." Nausea, vomiting, lightheadedness, dizziness, and fainting, are caused by orthostatic intolerance. Men and women both suffer from it, but women astronauts experience orthostatic intolerance more often than men.

If you've ever stood up after lying down for a while and felt dizzy, then you've experienced orthostatic intolerance. It's nothing to worry about, and it usually passes quickly. Researchers study orthostatic intolerance in astronauts because of concerns that it may affect an astronaut while landing a spacecraft. It could be disastrous if a Shuttle pilot were to pass out while trying to land. Fortunately, there are ways to help combat this problem.

"You have a complicated system of checks and balances involving the autonomic nervous system that will constrict your veins, ramp up your heart rate, etc. to accommodate changes in posture," Baker says. "But, we think some of those mechanisms get disrupted during space flight, especially the one that controls how your veins will constrict when you stand up so that you won't get blood pooling in your legs."

Wearing a G-suit reduces the pooling of blood in the legs by squeezing them with gas pressure. Also, drinking lots of water just before re-entry into the atmosphere helps replenish the amount of blood in the system. This allows the body to better support the elevation in blood pressure.

Before re-entry, Baker drinks water and takes salt tablets. "You want to get a mix that ends up being *isotonic*, about the same salinity as your blood, and that way you'll retain it longer in your venous system." Some astronauts don't like the taste of salt tablets and will drink chicken broth or a sports drink that contains electrolytes and salts. There are always a variety of these liquids onboard for the astronauts to choose from.

Bone Loss and Exercise

A one-week trip to space doesn't have much affect on bone density, but months in space do. Bone density loss is a huge concern among astronauts who spend a lot of time in space.

Women and men lose the same amount of bone while in space. The lack of gravity causes bones to lose calcium and become porous. This means they can break more easily.

A NASA study found that astronauts lose one percent of their bone mass per month! And that's with "existing countermeasures," including two hours of strenuous resistance exercise each day while in orbit. Male astronauts are just as much at risk for bone loss as women astronauts. So, finding a way to keep bones strong will benefit all space travelers.

Bones are living things and the calcium in bones is constantly being replaced by new calcium. Human beings need between 400 and 800 units of Vitamin D per day to help the body to absorb calcium. Depending on age, a person needs between 1000 and 1500 units of calcium per day to keep bones healthy. This is why it is so important to drink milk and eat foods that are rich in calcium.

Gravity actually encourages bones to keep replacing calcium. When astronauts aren't affected by gravity, their bones stop replacing as much calcium because the body doesn't need bones to survive in space. When bones start to lose calcium, both men and women astronauts are at risk for getting kidney stones and osteoporosis.

Millie Hughes-Fulford was a payload specialist on STS-40 *Columbia* in June 1991, the first Spacelab mission dedicated to biomedical experiments. Since then, Hughes-Fulford and her team of researchers at the University of California at San Francisco and the VA Medical Center in San Francisco have flown several cell biology payloads on the space shuttle. "Our theory is that the bone remodels with stimulation from exercise," says Hughes-Fulford. "So, what we think is happening is, there are mechanical stresses that are activated here on Earth that aren't activated in spaceflight."

Scientists don't know for sure what effect bone loss in space will have on men and women astronauts as they get older. But, these astronauts will be at a higher *risk* of getting the disease *osteoporosis*.

Osteoporosis literally means "porous bones." When this happens, bones lose minerals like calcium and may break easily. Both men and women can get osteoporosis. But as a woman ages and goes through menopause, osteoporosis becomes a much higher risk for her.

The loss of calcium during menopause puts a woman at a high risk of osteoporosis. Men are also at risk of osteoporosis as they get older, but women are four times more likely to suffer bone loss. In most cases, hormone replacement therapy can help relieve the symptoms of menopause and help to prevent further bone loss.

There is evidence that synthetic hormones can help to prevent osteoporosis in women of childbearing age. A recent study found that spinal bone density was 3.3 percent higher in women who used synthetic hormones. After five years of using synthetic hormones, these same women continued to increase their bone density at a rate of 0.2 percent per year. That's in contrast to the estimated 1.5 percent per year loss of bone density during menopause. In 10 years, a woman could lose 15 percent of her skeleton!

For postmenopausal women, hormone replacement therapy can't reverse osteoporosis completely, but it can help to build back some of the bone and slow the loss of more bone.

Scientists believe that the use of synthetic hormones by younger women astronauts will help keep up their bone density while on long space missions. Also, hormone replacement therapy will help postmenopausal astronauts prevent further bone loss.

Exercise and Weakness

It's extremely important for astronauts on long missions to exercise vigorously each day while in space. We use muscles every day just holding ourselves up against the Earth's gravity. Because living in space requires very little physical effort, and muscles tend to atrophy when not used, astronauts lose muscle if they don't exercise. They can experience weakness upon returning to Earth gravity.

"It's very critical that you exercise so that when you return to Earth you're not a total jellyfish," says Ellen Baker. "It's not as critical on the Shuttle flights, because a week or two will not really affect you that much. The day you return, even if you haven't exercised for two weeks, you might feel a little weak." Exercise is crucial, however, for long space missions to the International Space Station and beyond. "So if you don't want to exercise, you wouldn't want to go to the space station," says Baker.

Astronauts on the space shuttle use a treadmill to keep their muscles in shape. The treadmill is used with a restraint system to allow an astronaut to run or jog while in zero gravity. Force cords are attached at the waist and shoulders to tightly restrain the astronaut during exercise. These cords not only keep the astronaut from floating away, they act like gravity, ensuring the body has something to resist against. This resistance helps build muscle and stimulates bone growth.

Shannon Lucid uses the treadmill on the space station Mir

An electronic device monitors the astronaut's heart rate, run time, and distance. The heart rate is determined by an ear clip, which has an infrared sensor that detects increased blood flow in the ear lobe with each heartbeat.

Astronauts on the International Space Station exercise every day on a treadmill or stationary bicycle. "We do a combination of exercises. We have a bike that we use to pedal against resistance," says Baker. "And we have a resistive device so you can do simulated weight lifting, I guess you could say." Baker explains that the astronauts do a combination of aerobic and anaerobic exercises. Anaerobic exercise is when you exert yourself for short periods of time and then rest a while. It's a very effective way to build up muscle.

Space Medicine in the Future

There aren't enough astronauts to get good statistics on any medical issue, especially those that look at possible health issues between women and men. Science is based on gathering medical statistics, so with only a small statistical sample it's difficult to get meaningful scientific data.

Radiation, muscle atrophy, and bone loss affect both women and men and are today's biggest challenges in space medicine. More work must be done to determine what the health risks are for anyone who is planning to spend a long time in space.

Do Women Make Better Astronauts Than Men?

The short answer is "no." Women astronauts do have some advantages over male astronauts and vice versa. Women tend to be smaller, eat less, produce less waste, use less oxygen, etc. But men tend to have longer arms, which can be useful for some EVA activities. Baker agrees, "In my opinion, the differences between men and women are not as great as differences between individuals. So at this point in time, being that there is a relatively small number of people who have flown in space, I think the individual differences are the ones that stand out the most." It seems that neither gender has an advantage over the other.

A few years ago, rumors started circulating about an all-women space shuttle mission. What would be the purpose of such a mission? If biology, would the entire crew need to be women? NASA spokesperson Eileen Hawley says, "There are currently no plans to fly an all-female crew purely for the purpose of doing so. I don't doubt that at some point we may, but it will be in the natural progression of assigning crews based on qualifications and mission requirements."

Conclusion

False concerns about women's health issues are a thing of the past. Women Astronauts don't let health issues get in the way of their careers, or their lives. Both women and men who are planning to do any traveling in space need to keep their bones strong with the proper food and exercise.

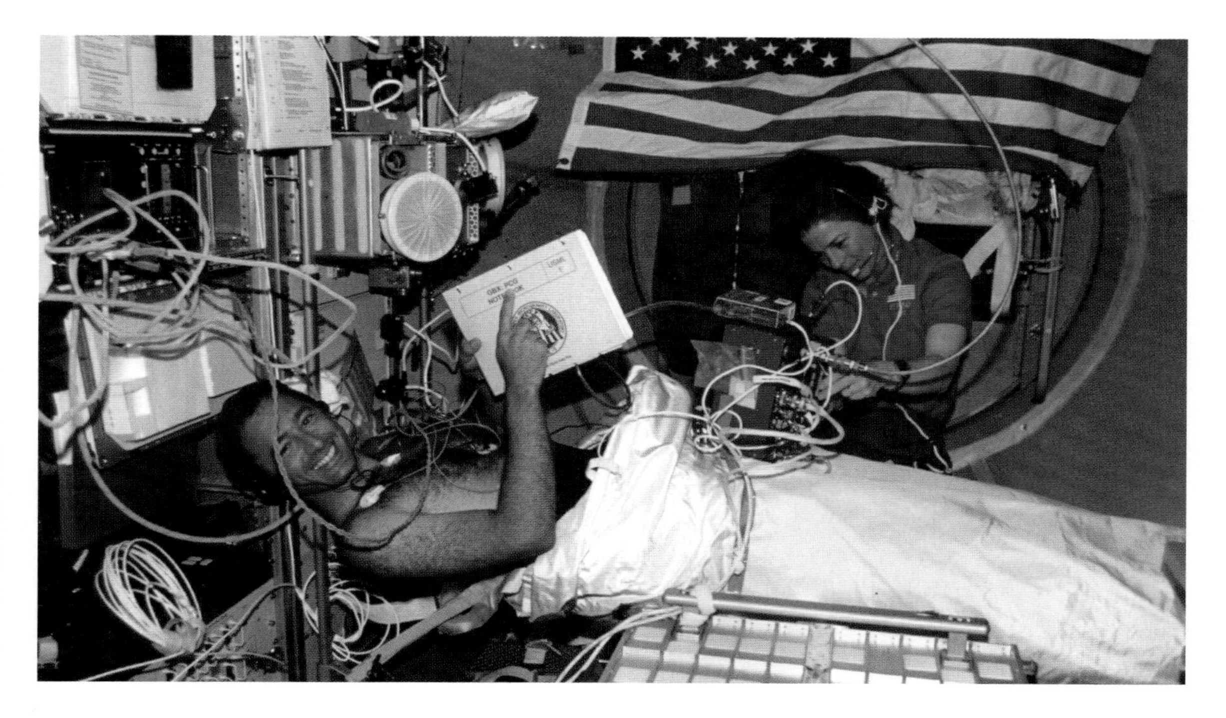

Bonnie Dunbar running medical tests on crewmate Lawrence DeLucas

Further research is needed to get a better picture of how the human body reacts to zero gravity. Over the coming decades, more women and men will become astronauts and travel into space, increasing our knowledge about the universe and the human body.

Women currently fill every position available to an astronaut and will continue to do so in the future. Becoming an astronaut is difficult for both women and men, but if you have the talent and the desire, you can do it. Gender-based health issues are not going to hold you back.

6

Becoming an Astronaut

Now that we've learned some space history, met the women astronauts, and discovered how to live in space and stay healthy, let's find out what it takes to become an astronaut, and how to learn more.

If you are interested in becoming an astronaut, here's where to start. The process takes a long time. It begins in school and continues even after astronaut candidates are chosen. *Remember, the women astronauts you've met in this book did it, and so can you.*

Choosing Astronauts

NASA's astronaut selection process begins when the space agency decides it needs more astronauts to work for them, typically every two years. The agency puts the word out on its web sites, in press releases, and to the news media. At this point, thousands of people fill out the standard government application forms, send them to Johnson Space Center's Astronaut Selection Office, and wait for a response. The Astronaut Selection Office then takes about a year to sort through the applications and invite the most qualified 100 or so people in for orientations, interviews and medical tests.

The world of space exploration has changed a lot since NASA chose its first *Mercury* astronauts in 1959. The first U.S. astronauts were all white men with military backgrounds. Today's astronauts are women and men of all ethnicities, and many professions.

"The first thing to recognize is that the field is wide open," says Sally Ride. "Over 20% of the astronaut corps is women. NASA is committed to having women be an increasing part of the corps." There are no quotas, but NASA encourages women and minorities to apply.

Friends Susan Helms and Janet Kavandi reunite on the ISS

There are a few different types of astronauts. Mission Specialists, Pilots, Payload Specialists, and Educator Mission Specialists each have different responsibilities and qualification requirements.

Mission Specialists are trained in the details of how the space shuttle works, as well as specific payloads and experiments. They are responsible for making sure the Shuttle systems are working correctly, planning the crew's activities, conducting space walks, doing experiments, and operating the remote manipulator system (RMS).

To become a Mission Specialist, applicants must have a bachelor's degree, and preferably an advanced degree, in a physical science, engineering, medicine, biology, or mathematics. They must also be able to pass a Class 2 physical, which includes vision correctable to 20/20, blood pressure of 140/90 or lower while sitting, and height between 58.5 and 76 inches.

Pilot astronauts can serve as space shuttle commanders and pilots. They are responsible for the overall safety and success of the mission, maneuvering the Shuttle during launch, while in space, and upon re-entry and landing. They may also help deploy and retrieve satellites by using the RMS.

In addition to the requirements for Mission Specialist, a Pilot applicant must have at least 1000 hours of jet flying experience, and able to pass a Class 1 physical. Blood pressure is the same as for the Mission Specialist position, however, a pilot's uncorrected vision must be 20/70 or better, and correctable to 20/20, and their height must be between 64 and 76 inches.

Payload Specialists are a special class of astronaut. They don't have to be NASA astronauts in order to fly on the space shuttle. Pilot and Mission Specialist candidates must be citizens of the United States, but Payload Specialists can be of any nationality. Payload Specialists have specific onboard duties for a particular payload, usually a satellite or experiment. Nominated by their employer, Payload Specialists must go through the same medical tests and training as the rest of their crewmates.

The Educator Mission Specialist position is a brand new type of astronaut. NASA plans to send teachers and other educators into space with this new program focused on teaching school kids about the wonders of the universe.

Members of the U.S. Military have an extra step in the astronaut selection process. "For example, if you are in the Navy, the Navy will send out a message saying that NASA is starting the next round of Astronaut selections," says Wendy Lawrence. "So I actually have to send my application first to the Navy selection board, and if I make it through them, they in turn send the application on to NASA. From that point, the process is exactly the same."

The key things that the Astronaut Selection Board (ASB) looks for are demonstrated technical expertise, the ability to learn new things, and the ability to work well in a team. The ASB invites the top one hundred or so applicants to NASA's Johnson Space Center (JSC) for a week of personal interviews and medical evaluations. After education, experience, and health have been considered, the personal interview is what really makes the difference. What makes a good astronaut is a self-confident, highly skilled generalist who works well in teams.

ASCAN: Astronaut Candidate Basic Training

The lucky few who are chosen, are invited to JSC in Houston, Texas for 1 to 2 years of astronaut basic training (ASCAN) and evaluation. When they arrive at JSC, they are considered "astronaut candidates." Those who make it through basic training and graduate to "astronaut" are expected to stay with NASA for at least 5 years.

"It really takes four years to have somebody ready to be assigned to a flight," says Anna Fisher. "You need those two years of ASCAN, then you need one or two jobs under your belt just to realize how NASA really works."

Astronaut candidates take classes on space shuttle systems, astronomy, navigation and guidance, orbital dynamics, meteorology, oceanography, geology, physics, mathematics, materials processing, as well as land and sea survival, scuba diving, and space suit training. They spend lots of time reading space shuttle manuals and taking computer-based lessons on everything from propulsion to environmental control.

Wendy Lawrence says that the hardest part of being an astronaut is, "the first year that you go through training as an astronaut candidate. Just assimilating all the information given out to you. The systems on the space shuttle aren't that much more complex than the systems on a military aircraft. It's just that there are more of them."

"Packing it all into your head and being able to recall it. It all boils down to information management. That's really the challenge," says Lawrence. "Astronaut candidates must figure out what is important to remember, "across the board on all the systems, and then the interactions between the systems and other aspects of flying the space shuttle."

Candidates train for emergencies in high and low atmospheric pressure chambers and during their first month of training, candidates must pass a swimming test. In a flight suit and tennis shoes, they must swim a 25-meter pool 3 times and must be able to tread water for 10 minutes.

Astronaut Julie Payette trains for water survival

Rhea Seddon and Ron McNair play in simulated weightlessness on a KC-135

Pilots must fly at least 15 hours per month in one of NASA's T-38 jets and practice Shuttle landings in the Shuttle Training Aircraft. Mission Specialist astronauts must fly at least 4 hours per month in a T-38 jet.

The "Vomit Comet" is the nickname given to a modified KC-135 airplane that NASA uses to give astronaut candidates a taste of weightlessness. The "Reduced-Gravity Aircraft Facility" can provide partial gravity to simulate the Moon and Mars. It does this by taking ballistic flight paths, like the motion of a ball hit by a baseball player. These paths are called parabolas. The airplane takes off, flies to a certain altitude, and takes a nosedive. For about 20 seconds, passengers experience weightlessness. This can be repeated up to 40 times in one day.

There is a reason it is called the "Vomit Comet' however. The feeling of microgravity causes some people to get space sick.

Watching films of astronauts on the KC-135 is like watching a bunch of little kids play. They do summersaults, fly like Superman, bump into each other by accident, and laugh a lot.

The Vomit Comet isn't typically used for heavy-duty astronaut training however. "I've been on a KC-135 exactly twice in my life," says Sally Ride. "And that's pretty common among astronauts. Everybody does it once." Other NASA employees use the plane for short microgravity experiments in biology and materials science.

Astronaut candidates learn about Shuttle operations on the Single Systems Trainer in the Space Vehicle Mockup Facility. Each astronaut has a trainer to help them work through checklists like on a real mission. The astronauts work on normal mission simulations and work through problems that could come up. All this training prepares astronauts for exactly what they need to do in an emergency.

The next level is the complex Shuttle Mission Simulators. These simulators train astronauts on how to work the shuttle during pre-launch, launch, orbit, re-entry and landing. An astronaut learns about manipulating the payload and maneuvering the Shuttle on a typical flight. When they are assigned to a flight, astronauts begin training on computerized simulators with specific mission plans.

Mission Specific Training

Pilots and Commanders train for Shuttle missions from launch to landing in motion-based simulators. Launch is simulated by rotating the cabin mockup by 90 degrees.

Millie Hughes-Fulford remembers that the simulations were very close to the real thing. "They only thing they didn't practice with us was when the solid rocket boosters come off. They were put together with explosive bolts, and we had never heard the sound before. So here we are going up and suddenly you hear 'ba-boom, ba-boom, ba-boom, ba-boom.' And it's like, 'what was that'? It got my attention. Then I didn't hear anybody say anything until someone said, 'there they go.' And I thought, oh that's what it was. But we never practiced that explosion. I had never heard it before."

Building 9 at the Johnson Space Center houses the Sonny Carter Training Facility Neutral Buoyancy Laboratory (NBL). It is the largest indoor pool in the world and is used to train astronauts to work in microgravity, space suits, and with hardware like the remote manipulator system. Water is a good simulator of microgravity because the weight of the astronaut and suit are balanced by air inside the suit. So, the astronaut floats in place.

Pilot astronauts train to land the shuttle at angles of 17 to 20 degrees at 300 miles per hour by using one of NASA's four modified Gulfstream II business jets. Pilots usually land this plane 600 times while training for a specific mission.

Between Missions

An astronaut's job starts on the ground. Astronauts are very busy people. Even those who are not training for a specific mission are working on all sorts of jobs to help other astronauts prepare for and do their jobs in space.

Astronauts at the Women's Forum at Kennedy Space Center

"When you come straight out of that ASCAN training," Anna Fisher says, "you have book knowledge and simulator knowledge, but you really don't realize how little you know in terms of how NASA works and functions."

Astronauts do a variety of jobs between missions. They review subsystems being built, write procedures, test hardware and software, work as Cape Crusaders, and attend public relations functions and space conferences. Most of the jobs are short-term with many hours and involve lots of meetings with other NASA workers and contractors.

"In between space flights we have what are called technical assignments," says Wendy Lawrence. "They are basically jobs that we have in the Astronaut Office to support both the space shuttle and space station programs. Currently I have a couple of technical assignments."

Some astronauts do CAPCOM duty in Mission Control. The CAPCOM is the person who talks with the crew and gives them instructions during their mission. Educator Mission Specialist Barbara Morgan is currently a CAPCOM for the International Space Station.

"It's extremely helpful to do a stint as a CAPCOM," says Anna Fisher. "I would have felt very uncomfortable going on a flight and never having worked in Mission Control. It's as important as any other training to understand how they make their decisions and how things are done. It gives you confidence in the things they do by having been a part of that team."

Being an Astronaut takes a lot of time. The biggest complaint that astronauts seem to have is that they don't have enough time to spend with their family and friends. Astronauts travel a lot and the heaviest time for travel begins three months before launch and ends two months after landing.

It's a lot of hard work, but astronauts enjoy hard work. Ellen Baker loves, "coming to work and having different things to do every day, I think that makes life interesting. I also like having challenges, problems that you can solve, and of course we get to do that every day. I also like working with a group of people who are motivated, focused, and interesting. I'm fortunate to be surrounded by people like that. And when I actually do get to go in space, there's nothing really more exciting than that. So it's fun, and you can work really hard. But you can have lots of fun while you are doing it."

How Much Money Does an Astronaut Make?

Astronauts are paid as civil service employees of the U.S. government. The pay depends on the grade and the number of years of service. In 1976, at the beginning of the Shuttle era, the job of astronaut paid between $11,000 and $34,000 per year depending on how much experience the astronaut had in their field. In 2002, an astronaut's basic salary ranges from about $40,000 to $107,000 per year.

Astronauts in the military continue to earn military pay instead of civil pay. But, depending on rank and the number of years in the service, it's about the same.

Can I Become an Astronaut?

So you want to be an Astronaut? If you're interested in a career in space exploration, ask yourself these questions: Do you enjoy math, science, technology and working with computers? Are you curious about the world around you? Are you curious why things work? Do you enjoy learning? Astronauts enjoy doing these things too.

How Can I Feel Like an Astronaut Now?

At this point, you must be wondering how long it will be before you get to blast off into orbit. Well, you don't have to wait. You can experience space today. Here are a few things that will make you *feel* like an astronaut. Now!

Whether you are young or old, a girl or a boy, there are lots of ways you can get involved in space exploration.

Go See an IMAX Space Movie

Get your ticket for an adventure in space. The IMAX Corporation has made several space movies that play on gigantic 6-story high movie screens. "*Those Imax movies are great,*" says Sally Ride. "They give by far the best feeling of what it's like to be in space."

The first was "Hail *Columbia*" which premiered in 1982. It was followed by "The Dream is Alive" (1985), "Blue Planet" (1990), "Destiny in Space" (1994), and "Mission to Mir" (1996).

The latest IMAX space movie is called "Space Station 3-D." This trip to the ISS opened in spring 2002 at IMAX theatres around the world. The movie chronicles the on-orbit assembly of the International Space Station. Viewers blast off on the Shuttle from Kennedy Space Center in Florida, experience floating in space, and a launch from Russia's Baikonur Cosmodrome. IMAX calls *Space Station 3-D* a, "home movie from humanity's home-away-from-home."

A view of the ISS from the Shuttle's payload bay, courtesy IMAX

Twenty-five astronauts learned how to use the IMAX cameras and how to direct and shoot a scene in zero-g. Two cameras were used to film this movie. One stayed on the ISS in orbit 220 miles above Earth, and one traveled to space and back on three different Shuttle missions. The ISS camera orbited the Earth for 337 days between September 2000 and August 2001. Bolted into the cargo bay of the space shuttle, the IMAX camera held a roll of film one mile long.

The IMAX Corporation plans to make its next big space movie using images beamed back to Earth by NASA's unmanned Mars Exploration Rovers. The producers promise the audience will feel like they are walking on the surface of the red planet. Filming starts in early 2004 when the rovers arrive on Mars.

Imax.com has information on theatre locations, movie schedules and ticket prices (tickets are usually $7 or less. Most science centers offer a discount if you show them your student I.D. card).

Visit Your Local Science Museum.

Many of today's science museums, exploratoriums, and learning centers are hands-on so that you can play with the exhibits. Even the most mature adults turn into little kids around these!

Planetariums allow you to see the night sky without city lights getting in your way. Most have changing multimedia presentations on space and astronomy, plus guest speakers.

Space explorers use all types of science and engineering to make space travel possible. When you visit your local planetarium, natural history museum, or technology center, you are learning about some of the technologies and sciences that people use to explore space.

Most Science Centers have a membership program through which you can get discounted or free admission. One such program in the U.S. is the Association of Science Technology Centers (ASTC) Passport Program. When you join, you get free admission to more than 200 science centers across the United States. More than 550 science centers, science museums, and related organizations in 43 countries make up the ASTC. Find out more at www.astc.org.

The National Air and Space Museum (NASM) in Washington D.C. is a great place to see American and Russian rockets. The museum has everything from the Wright Flyer to space suits worn by Cosmonaut Yuri Gagarin and NASA's John Glenn, and *Apollo* lunar suit space suits. You can even walk through the backup to the *Skylab* space station. In addition to exhibits, NASM has a planetarium and an IMAX theatre. If you can't visit the museum in person, the NASM has an excellent web site with many exhibits online at www.nasm.edu.

The U.S. Space & Rocket Center in Huntsville Alabama is the official visitor center for NASA's Marshall Space Flight Center. In addition to Space Camp, the center has a museum, an IMAX theatre, a gigantic Saturn V rocket, and more. They have a simulator called Space Shot that promises that you will, "feel four G's of liftoff force and the sensation of weightlessness."

Ah, the Space Gift Shop

It's always a good idea to bring some extra cash with you when visiting your local science museum because the gift store is always packed with tons of cool stuff. In addition to space and science toys, they have lots of educational materials like books, videos, and science kits.

The National Air and Space Museum has a three-story gift shop, the biggest of the Smithsonian museums, with all kinds of flying and space products. There is an entire wall of t-shirts, NASA logo stuff, space collector's items, and the original model of the Starship *Enterprise* from the first Star Trek T.V. Series.

If you can't get to a space museum, check out the web for space goodies. Dayna Steele Justiz, President of thespacestore.com and Space Store in Houston, Texas, says, "We do everything! We've got toys, mission patches, books, videos. We've got everything you could imagine."

Space Camp

Attending Space Camp is a great way to learn about what astronauts do and experience what it's like to fly on the space shuttle. Whether you are 9 years old or an adult, there are lots of great space activities to do at Space Camp. You can go through basic training like a real astronaut, fly on a virtual shuttle mission on the simulators, and get an idea of what it would be like to walk on the Moon in the 1/6th gravity chair. You also get to eat like the astronauts and sleep in space bunks. There are several programs for kids, adults and educators to choose from like the 5-day mission, Space Academy and the Weekend Aviation Challenge.

There are two Space Camp sites in the U.S. There are centers at the U.S. Space & Rocket Center in Huntsville, Alabama near NASA's Marshall Space Flight Center, and in Titusville, Florida near the Kennedy Space Center. Each Space Camp has programs for young people and adults too. The Huntsville Space Camp even has corporate getaways for companies. There are International Space Camps in Japan, Belgium, Turkey, and Canada. See the reference section for web addresses.

There are other science camps around the world for all ages, teachers, and families. The California Science Center http://www.casciencectr.org/ in Los Angeles has Science Camp every summer. They give classes and workshops and there is even a camping trip to Catalina Island, off the coast of Southern California.

The Sally Ride Science Club

Sally Ride started a company called Imaginary Lines, which runs the Sally Ride Science Club. "We do things like put on events that girls in upper elementary and middle school think are cool and want to come to with their friends," says Ride. "It keeps them engaged and in the pipeline for math and science careers."

The club is for girls interested in science, math, and technology. There is a members-only web site and newsletter, contests, and science festivals around the U.S. You can find out more about it at www.SallyRideClub.com. Adults will be able to join the club in Summer 2002.

"The idea is to give them a community to belong to of girls around the country that are also interested in science and math and computers," says Ride. "It gives them a way to talk with each other and with women who are involved in math and science careers; veterinarians, computer programmers, aerospace engineers, Mars scientists, etc."

There is also a special Space Camp program for parents and daughters called the Sally Ride Science Club Space Camp. It's for girls age 7 to 11. The adult can be a family member or friend age 19 or over. The idea behind this special program is to show girls the future they can have in math, science and technology.

Ride wants girls to see that they can have an exciting career in math, science, and technology. "There are a lot of people that are fascinated by careers in science and engineering," says Ride. "And these doors are wide open to them if they are interested."

The Girl Scouts

Knowing how to work in teams and work towards a goal is an essential part of being an astronaut. In fact, these are useful skills in anything that you do. The Girl Scouts has lots of fun and interesting activities that teach these skills, including an aerospace badge.

The Girl Scouts Aerospace Badge

Several women astronauts were Girl Scouts when they were young. "I loved Girl Scouts," says Millie Hughes-Fulford, who was in the group for almost 6 years. "I'm a big believer in Girl Scouts. It's good. It teaches girls how to do things, and work together."

Jan Davis has been a Girl Scout and an assistant troop leader. On her first mission in 1992, Davis brought "quite a few" yummy Girl Scout cookies onboard. The food technicians packaged them, "so I could bring them back and give them out to people. But I broke some out for a picture."

Other Girl Scouts-turned-astronauts are Anna Fisher, Wendy Lawrence, Janice Voss, Yvonne Cagle, Laurel Clark, Eileen Collins, Nancy Currie, Susan Helms, Joan Higginbotham, Kathryn Hire, Tamara Jernigan, Sandra Magnus, Barbara Morgan, Lisa Nowak, Rhea Seddon, Susan Still-Kilrain, Kathy Sullivan, Kathryn Thornton, and Mary Ellen Weber.

Visit a Space Center

At Kennedy Space Center's Visitor Complex in Florida, you can watch a launch, visit the Astronaut Hall of Fame, and walk inside a replica of the space shuttle.

NASA has space centers around the U.S. They all offer tours to groups. Gather your scout troop, science club, or classroom, and go! The space centers are Ames Research Center, Dryden Flight Research Center, Glenn Research Center, Goddard Space Flight Center, the Jet Propulsion Laboratory, Johnson Space Center, Kennedy Space Center, Langley Research Center, Marshall Space Flight Center, Stennis Space Center and Wallops Flight Facility.

Jan Davis eating Girl Scout cookies on the Shuttle

Check out www.nasa.gov for the web sites and locations of each center.

A Taste of Space

The freeze-dried astronaut ice cream that you can buy here on Earth isn't really what the astronauts eat in space. They carry up regular Earth ice cream in a medical freezer and usually eat it up soon after getting into space.

Eat like an astronaut. Dayna Steele Justiz says, "We're the only place in the world that can buy real space food. We sell the actual thing they send up." The Space Store sells all kinds of space foods manufactured by SpaceHab, the same company that makes and packages food for the space shuttle and ISS astronauts.

Space Groups

Learn as much as you can about space exploration and the sciences and technologies that make it possible. Here are a few suggestions. Check out the reference section of this book or the CD Rom as well as my web site www.Woodmansee.com/WomenAstronauts for more space organization links and information. Learning is fun and addictive. Join a group and learn skills that will make you better at whatever you choose to do.

Young Astronauts is the largest aerospace group for young people in the entire world. There are local chapters all over the world, a Space School program, and Young Astronauts on-line. The Young Astronauts web site is www.yac.org/yac.

Yuri's Night commemorates Yuri Gagarin's first flight into space. The organization puts on space parties all over the world each year on the 12th of April. The Yuri's Night web site is www.YurisNight.net.

The Women of Space organization (WoS) was formed in 2001 to celebrate and encourage the accomplishments of women in space exploration. The organization plans to have its first conference in Spring 2003 in Los Angeles. WoS also runs a speaker's bureau where you can request women who work in space exploration to come and talk with your group. The Women of Space web site is www.WomenOfSpace.org.

The International Space University is geared towards students who have already completed an undergraduate degree. ISU has a 2-month summer session program and an 11-month Master of Space Studies program. Students can focus on one of nine areas of interest including everything from space engineering, to life science, to space business. The ISU web site is www.isunet.edu.

Tourists in Space

Space the final tourist trap! Right now, people pay millions of dollars to go into space. But, that's beginning to change. Someday the average person may be able to buy a ticket to space.

"I think space tourism is a good idea because the more people who can experience the thing that I've had the chance to experience, the better," says Sally Ride. "It's just a wonderful experience."

A new spacecraft is being designed especially for space tourists. The Space Adventures 2-tourist craft, will go up to 62 miles (space starts at 50 miles), let you feel microgravity for 3 minutes, and you come back with the title of "astronaut." The ride is similar to what Alan Shepard experienced on his first flight into space. The company expects rides to cost $100,000 per person. Space Adventures expects the price will drop as more people buy tickets to ride to the edge of space.

If you work hard and save up about $5,000, you can fly in the Russian version of a KC-135, otherwise known as the vomit comet, and experience weightlessness for 20 seconds at a time as the airplane flies in parabolas.

Advice from Women Astronauts

"Getting a good education will open doors for you,' says Wendy Lawrence. "As trite as it may sound, it really is true. The advice I give to anybody, young boy or girl alike, is to find something that you really enjoy doing and then do it to the best of your ability. Be willing to pursue your dream. Dreams can come true, but recognize the fact that if you are going to pursue your dream, that means that you're going to have to work hard. The dream is not going to be given to you. There are going to be ups and downs. So if you really want to see something come true and you really want to be able to do something in your life, you have to accept the responsibility of working hard to achieve it."

Sally Ride's advice is, "Find the area in science and engineering that most interests you, and then pursue that as far as you can. NASA doesn't care what the area of science or engineering is, they just want demonstration of commitment to that area."

Sally Ride on the Shuttle's mid deck on her first flight, STS-7

Millie Hughes-Fulford says, "I just think that young girls, or young boys should look around the world and see what looks interesting. Use that as your model. And read everything you can about it, focus on it, do internships, get as much experience as possible, seek out mentors. That's something girls don't do as much as boys. Look for a mentor. Get someone who can help you. Get someone you can call and get advice from. Seek out people in your field that can help you, that will be knowledgeable, and can tell you what direction you should take."

Jan Davis's advice is, "You should select whichever field interests you. In other words don't just pick one over the other because you think you'd have a better chance if you picked, say engineering over being a veterinarian. If you want to be a veterinarian you can be an astronaut. We have one. Or if you want to go military or non-military, choose what you think is best for you. Choose what you would be the most interested in and do the very best you can!"

"Study math and science," advises Janice Voss. "Do what you enjoy and are good at, because then you will excel."

Heidemarie Stefanyshyn-Piper's advice is to, "Study hard in school, and always remember that math and science are fun! Pick a career field that you enjoy doing and then do the best job that you can."

Ellen Baker says, *"Choose something you really like because the odds of getting into the Astronaut Office are pretty small.* If you don't get into the Astronaut Office, you don't want to be stuck with a degree you're not interested in. One of the common threads here is people do their best and you always have to be proud of what you do," says Baker. "You never know what's going to happen in 10 years, perhaps the needs of the Astronaut Office will change, or we'll be sending people with different skills into space."

"You need to know what your interests are," says Shannon Lucid. "And along that line, you may not know all your interests. You need to try different things and take advantage of every opportunity that comes your way, so that you'll be able to find out what you're really interested in." She goes on to say, "You can't control whether you become an astronaut or not. When things outside of your control don't go the way you want them, you can't think, 'Oh my life is ruined.' Make the most of the things you can control and enjoy everyday that you have."

Shannon Lucid and John Blaha are smiling because Lucid has just spent her 752nd hour in space, a space shuttle record.

Conclusion

Today's astronauts come from several disciplines, most are still technical. There are military pilots, chemists, biologists, medical and veterinary doctors, engineers, physicists, geologists, astronomers, and more.

In the future, as more people are needed to work in space, teachers, journalists, artists, and even entertainers will be needed to bring the experience of space flight to people around the world.

The bottom line is, if you really want to be an astronaut, get a good education and stay in good physical shape. It also helps to learn how to fly an airplane, but if you don't, you will get a chance when you enter the astronaut program.

Astronauts make up a small part of the space exploration workforce. You don't have to be a scientist or engineer to get involved in space exploration. There are many other jobs supporting astronauts and other careers in space exploration too!

All kinds of people, with all kinds of skills, work to support human space exploration. There are engineers, biologists and doctors, scientists, computer programmers, technicians, test technicians, safety inspectors, managers, communications experts, science writers, technical writers, photographers, public affairs people, contract specialists, budget accountants, and dozens more space exploration workers.

"There are other really great jobs at NASA," says Shannon Lucid. "It's really important not to get so focused on thinking, 'ok I've got to get this one great job,' and don't do anything else in life. You have to look at life as a total."

The people on the ground make the technical accomplishments that happen in space possible. Talk to any of these space workers and you will find that they take great pride in their work. In a way, they go into space on each and every mission. Many astronauts were space workers before they became astronauts.

Understanding science is important for everyone. No matter what they do. Science is fun and learning is fun. *Astronaut's say it's fun to be a science geek because the stuff you get to learn about is so cool!*

In the reference section of this book are more places to learn about space and astronauts. Join a group of space explorers in your local area. If you are looking into colleges, remember there are many student loans and scholarships for women and minorities.

Be brave and study hard. Learn all you can about the things that you are interested in. They will take you to the planets and beyond.

Conclusion

Our Future in Space

All of the women that you've met in this book are passionate about space and hopefully you've come away with some of their excitement. I hope that you've enjoyed reading *Women Astronauts* as much as I've enjoyed researching, learning, talking with the astronauts, and relating it all to you.

Whatever you choose to do in life, it's important that you do something you believe in! You can do whatever you want to do. Don't listen to anyone who tells you that you can't. You can achieve your dreams, just like the women in this book.

A good education is the key to your success. Not just in school, but learning on your own about whatever interests you. Be curious. Be a lifetime learner. Find something that you are passionate about and do it!

You can help shape the future in whatever you do. If you care about space exploration you can get involved and make a difference. Even if you don't become an astronaut, there are still ways to reach space.

For example, as a science writer, I get to learn about space exploration, talk with astronauts, scientists, engineers and others, and share that knowledge with people like you. Space artists show us worlds that we might never see, and inspire others to make it a reality. Engineers, doctors, and many others advance us towards the stars. Space exploration is about working together for a common goal.

Space is for everyone, not just those who are lucky enough to become astronauts. We can all experience space travel through our imagination and with the stories, pictures, and impressions brought back to Earth. More of us will go into space as our technology advances. It's just a matter of time before space tourism becomes affordable for the average person. You could be one of those who help to make this happen.

Exploration is important to the future of humanity, as it always has been. Now that most of the landmass of the Earth has been explored, it's important that we explore beyond our planet. If we don't venture into space, then the human race will always be confined to this planet, we will know nothing about the universe other than what we can see out our own window. That's not what we want! We want to travel into space and experience the wonder of the universe for ourselves.

With international partners working together, instead of against each other, to explore the cosmos, we can improve the lives of everyone on this planet. The ecology of our home planet will survive better once we learn more about the universe around us.

Our future is in space and someday soon, astronauts will travel to Mars and to other planets in our solar system. In the distant future, human beings will travel to worlds around distant stars. *You can help create this future!*

Space exploration is a long-term enterprise. It takes money and it takes political will. We can make a difference by telling our political leaders that space is important to us. We need to support science and technology in many fields, and in every country, to make this happen. We cannot be shortsighted and only look to the here and now. Space holds the key to our long-term survival.

That we can perceive and understand our universe is a gift. That we can dream of a better future is an even greater gift. We must stay true to our dreams and look beyond the problems of today to create a brighter future tomorrow. I believe that our future is in space and I hope that you do too.

So, I hope that you've enjoyed learning about our pioneering women astronauts. Whatever you choose to do in life, I wish you all the happiness in the universe. Follow your heart. Follow your dreams. And you may find yourself among the stars.

Appendix A:

Selected Bibliography for *Women Astronauts*

BOOKS

These books and others are available through my web site at www.woodmansee.com/WomenAstronauts.

Title: **I Want to be an Astronaut**
Author: Stephanie Maze
Date published: March 1999
Reading level: Juvenile Literature, ages 9-12
Publisher: Harcourt

Title: **Women in Space**
Author: Carole S. Briggs
Date published: 1999 (revised edition)
Reading level: Juvenile Literature
Publisher: Lerner Publications Company (www.lernerbooks.com)

Title: **To Space and Back**
Author: Sally Ride with Susan Okie
Date published: 1986
Reading level: Juvenile Literature, ages 4-8
Publisher: Lothrop Lee & Shepard

Title: **Eyewitness Books/ Astronaut: Living in Space**
Author: Kate Hayden
Date published: 2000
Reading level: Juvenile Literature, grades 1-3
Publisher: Dorling Kindersley Limited

Title: **Outer Space Directory**
Author: Spencer Kope
Date published: 1997
Reading level: all
Publisher: Willow Creek Press of Washington

Title: **Space and the American Imagination**
Author: Howard McCurdy
Date published: 1997
Reading level: adult
Publisher: Smithsonian History of Aviation Series

Title: **The Right Stuff, the Wrong Sex: The Lovelace Woman in Space Program** (thesis)
Author: Margaret Weitekamp
Date published: 2000
Reading level: Adult

Title: **The Astronaut Fact Book**
Author: NASA
Date published: September 2000
Publisher: NASA

Title: **Who's Who in Space: The First 25 Years**
Author: Michael Cassuttt
Date published: 1987
Reading level: Adult
Publisher: G. &K. Hall

Title: **Cool Careers for Girls in Air and Space**
Authors: Ceel Pasternak & Linda Thornberg
Date published: 2001
Reading level: Juvenile Literature
Publisher: Impact Publications

Title: **The Complete Idiot's Guide to NASA**
Authors: Thomas D. Jones & Michael Benson
Date Published: 2002
Reading level: all
Publisher: Alpha Books

WEB SITES

Title: The **NASA** web sites:
UTL: http://www.nasa.gov
Start here for your search of the NASA web.

Title: **European Space Agency (ESA)**
URL: www.esa.int

Title: **Russian Space Research Institute**
URL: http://arc.iki.rssi.ru/eng/index.htm

Title: **Japan Space Agency (NASDA)**
URL: yyy.tksc.nasda.go.jp

Title: **The Canadian Space Agency**
URL: http://www.space.gc.ca/home/index.asp

Appendix B:

Resources

In addition to the selected bibliography, which has great resources that I used, here are some other resources you might want to use.

BOOKS

Astronaut: Living in Space by Kate Hayden

Cosmos by Carl Sagan

Floating in Space by Franklyn M. Branley, Harper Collins, 1998

National Geographic's *Our Universe* by Roy A. Gallant

The Right Stuff by Tom Wolfe

The Right Stuff, the Wrong Sex by Margaret Weitekamp (available in 2003)

Space Exploration by Eyewitness Books

Your Future in Space: The U.S. Space Camp Training Program by Flip and Debra Schulke and Penelope and Raymond McPhee, Crown Publishers Inc. NY, 1986.

WEB SITES

Title: **Women Astronauts Book**
URL:
http://www.woodmansee.com/womenastronauts
This is my web site for the book, which I will update with new resources and information

Space Agencies

Title: **NASA Web Home page**
URL: http://www.nasa.gov

Title: **European Space Agency (ESA)**
URL: http://www.esa.int/export/esaCP/index.html

Title: **The Canadian Space Agency**
URL: http://www.space.gc.ca/home/index.asp

Title: **The National Space Development Agency of Japan (NASDA)**
URL: http://www.nasda.go.jp/index_e.html

Title: **Russian Space Science Internet**
URL: http://www.rssi.ru/

Clubs and Groups

Title: **The Sally Ride Science Club**
URL: www.SallyRideClub.com.

Title: **Young Astronauts Corps**
URL: http://www.yac.org/yac

Title: **Girls Scouts USA**
URL: http://www.girlscouts.org/

Title: **World Association of Girl Scouts and Girl Guides**
URL: http://www.wagggsworld.org

Title: **Women of Space**
URL: http://www.womenofspace.org/

Title: **Yuri's Night**
URL: http://www.yurisnight.net/

Title: **International Space University**
URL: www.isunet.edu

Title: **The Planetary Society**
URL: www.planetary.org

Places to Go

Title: **The Space Exploration Museum Database Page** (for a list of some places to go)
URL:
http://www.cyberspacemuseum.com/spacedbase.html

Title: **National Air and Space Museum**
URL: http://www.nasm.si.edu/

Title: **The Challenger Center**
URL: http://www.challenger.org/

Title: **McAuliffe Challenger Center at Framingham State College**
URL: http://www.christa.org/

Title: **Center of Science and Industry (COSI)**
URL: http://cosi.org

Title: **California Science Center**
URL: http://www.casciencectr.org/

Title: **The San Francisco Exploratorium**
URL: http://www.exploratorium.edu/

Title: **The Rose Center for Earth and Space**
URL: www.amnh.org/rose

Title: **Space Camp (by Nationality)**
URL: USA: http://www.spacecamp.com
URL: Canada: http://www.spacecampcanada.com
URL: Japan: http://www.spaceworld.co.jp
URL: Belgium: http://www.ping.be/eurospace
URL: Turkey: http://www.spacecampturkey.com

Title: **Space Adventures**
URL: http://www.spaceadventures.com/index.html

Sites for Kids

Title: **So You Want to be an Astronaut**
URL:
http://liftoff.msfc.nasa.gov/academy/astronauts/wannabe.html

Title: **The Starchild Learning Center for Young Astronauts**
URL: http://starchild.gsfc.nasa.gov

Title: **Space Kids**
URL: http://spacekids.hq.nasa.gov/

Space Images

Title: **Earth from Space**
URL: http://earth.jsc.nasa.gov/

Title: **Great Images in NASA (GRIN)**
URL: http://grin.hq.nasa.gov/subject-space.html

Title: **Astronomy Picture of the Day**
URL: http://antwrp.gsfc.nasa.gov/apod/astropix.html

Sites on Women Astronauts

Title: **Women in Microgravity**
URL: http://microgravity.msfc.nasa.gov/WOMEN/

Title: **Women of NASA**
URL:
http://questdb.arc.nasa.gov/content_search_women.htm

Sites with other Space Information

Title: **The Astronaut Connection**
URL: http://www.nauts.com

Title: **The Nine Planets**
URL: http://www.seds.org/nineplanets/nineplanets/

Title: **Encyclopedia Astronautica**
URL: http://www.Astronautix.com

Title: **Astronaut Application forms**
URL:
http://www.nasajobs.nasa.gov/jobs/astronauts/aso/application.htm

Space News

Title: **Space Daily**
URL: http://www.spacedaily.com

Title: **Space.Com**
URL: http://www.space.com

Places to get Space Stuff

Title: **The Space Store**
URL: http://www.thespacestore.com/

Title: **Collect Space**
URL: http://www.collectspace.com/

Appendix C:

Table of Women Astronauts

Name	Nationality	1st Mission	Launch Date	Landing Date	Other Missions	Comments	Position
Valentina Tereshkova	USSR	Vostok 6	16-Jun-63	19-Jun-63	None	1st woman in space, orbited 48 times	Cosmonaut
Svetlana Savitskaya	USSR	Soyuz T-7	19-Aug-82	27-Aug-82	Soyuz T-12: 7-17-84 to 7-29-84	1st woman EVA in July 1984 on Soyuz T-12, 1st woman to fly in space twice.	Cosmonaut
Sally Ride	US	STS-7 Challenger	18-Jun-83	24-Jun-83	STS-41G	1st American woman in space, 1st US woman to fly in space twice.	Mission Specialist
Judith Resnik	US	41-D Discovery	30-Aug-84	5-Sep-84	STS-51-L	Killed on 1-28-96 in Challenger explosion	Mission Specialist
Kathryn Sullivan	US	41-G Challenger	5-Oct-84	13-Oct-84	STS-31, STS-45	1st US woman EVA. 1st mission to have 2 women, 3.5 hour EVA on STS-41-G (1st Woman PC in 1992)	Mission Specialist
Anna Fisher	US	51-A Discovery	8-Nov-84	16-Nov-84	None	First mom in space	Mission Specialist
Margaret Rhea Seddon	US	51-D Discovery	12-Apr-85	19-Apr-85	STS-40, STS-58	Figured out how to do CPR in zero-g.	Mission Specialist on 51-D & 40. Payload Specialist 58.
Shannon Lucid	US	51-G Discovery	17-Jun-85	24-Jun-85	STS-34, STS-43, STS-58, up on STS-76, down on STS-79	First American Woman to Spend Extended Time on Russia's Mir Space Station: 188 days	Mission Specialist on 51-g, 34, 43, 58. Board Engineer on Mir.
Bonnie Dunbar	US	61-A Challenger	30-Oct-85	6-Nov-85	STS-32, STS-50, STS-71, STS-89		Mission Specialist on 61-a, 32, 71. Payload Commander on 50, 89.
Mary L. Cleave	US	61-B Atlantis	26-Nov-85	3-Dec-85	STS-30		Mission Specialist on 61-b, 30
Sharon Christa McAuliffe	US	51-L Challenger	28-Jan-86	N/A	N/A	Killed on 1-28-96 in Challenger explosion	Teacher
Ellen S. Baker	US	STS-34 Atlantis	18-Oct-89	23-Oct-89	STS-50, STS-71		Mission Specialist
Kathryn C. Thornton	US	STS-33 Discovery	22-Nov-89	27-Nov-89	STS-41, STS-61	21 hrs of EVA on 49 and 61	Mission Specialist. Payload Cdr on 73
Marsha S. Ivins	US	STS-32 Columbia	9-Jan-90	20-Jan-90	STS-46, STS-62, STS-81, STS-98		Mission Specialist. Also flight eng on 98
Linda M. Godwin	US	STS-37 Atlantis	5-Apr-91	11-Apr-91	STS-59, STS-76, STS-108	EVA on 76, and 108	Mission Specialist on 37, 76, 108. Payload Cdr on 59.
Helen Patricia Sharman (UK/Soyuz)	UK	Soviet Soyuz TM-12	18-May-91	26-May-91	None	First British citizen in space	Tourist
Millie Hughes-Fulford	US	STS-40 Columbia	5-Jun-91	14-Jun-91	None		Payload Specialist
Tamara E. Jernigan	US	STS-40 Columbia	5-Jun-91	14-Jun-91	STS-52, STS-67, STS-80, STS-96	STS-96 EVA married to astronaut Peter J. "Jeff" Wisoff. Never flew together.	Mission Specialist on 40, 52, 80, 96. Payload Cdr on 67.
Roberta L. Bondar	Canada	STS-42 Discovery	22-Jan-92	30-Jan-92	None	1st Canadian woman	Payload Specialist
N. Jan Davis	US	STS-47 Endeavour	12-Sep-92	20-Sep-92	STS-60, STS-85	Married and astronaut Mark Lee, flew with him on STS-47, 1st married couple in space, met during astronaut training. Later divorced.	Mission Specialist on 47, 60. Payload Cdr on 85.

Mae Jemison	US	STS-47 Endeavour	12-Sep-92	20-Sep-92	None	1st African-American woman astronaut	Science mission Specialist
Susan J. Helms	US	STS-54 Endeavour	13-Jan-93	19-Jan-93	STS-64, STS-78, STS-101, up on STS-102, down on STS-105	1st woman ISS resident, EVA on STS-102	Mission Specialist. Payload Cdr on 78. Flight Eng on 64.
Ellen Ochoa	US	STS-56 Discovery	8-Apr-93	17-Apr-93	STS-66, STS-96	1st Hispanic woman astronaut	M.S on 56. P.C. on 66. M.S. and F.E. on 96.
Nancy J. (Sherlock) Currie	US	STS-57 Endeavour	21-Jun-93	1-Jul-93	STS-70, STS-88, STS-109 in 2002		Mission Specialist
Janice E. Voss	US	STS-57 Endeavour	21-Jun-93	1-Jul-93	STS-63, STS-83, STS-94, STS-99		MS on 57, 63, 99. PC on 83, 94.
Chiaki Naito-Mukai	Japan	STS-65 Columbia	8-Jul-94	23-Jul-94	STS-95	Japan's 1st woman astronaut	Payload Specialist
Elena V. Kondakova	Russian	Soyuz TM-17	4-Oct-94	22-Mar-95	STS-84 Atlantis	169 days on Mir, 1st woman (& Russian woman) resident on Mir	Mission Specialist
Eileen Marie Collins	US	STS-63 Discovery	3-Feb-95	11-Feb-95	STS-84, STS-93	1st female pilot and commander	Pilot on 63, 84. Cdr. On 93.
Wendy B. Lawrence	US	STS-67 Endeavour	2-Mar-95	18-Mar-95	STS-86, STS-91		Mission Specialist
Mary Ellen Webber	US	STS-70 Discovery	13-Jul-95	22-Jul-95	STS-101 Atlantis		Mission Specialist
Cathryn G. Coleman	US	STS-73 Columbia	20-Oct-95	5-Nov-95	STS-93		Mission Specialist
Claudie (Andre-DeShays) Haigneré	France	Russian-French Cassiopée Mir mission on Soyuz TM-24	17-Aug-96	2-Sep-96	Soyuz TM-33 - - 2001 Oct 21	Only woman in current ESA astronaut corps	Research Cosmonaut
Susan L. (Still) Kilrain	US	STS-83 Columbia	4-Apr-97	8-Apr-97	STS-94		Pilot
Kalpana Chawla	US	STS-87 Columbia	19-Nov-97	5-Dec-97	Planned for 2002: STS-107	1st Indian-born Woman in space	Mission Specialist
Kathryn Hire	US	STS-90 Columbia	17-Apr-98	3-May-98	None		Mission Specialist
Janet Kavandi	US	STS-91 Discovery	2-Jun-98	12-Jun-98	STS-99, STS-104 ISS assembly flight		Mission Specialist
Julie Payette	Canada	STS-96 Discovery	27-May-99	6-Jun-99		1st Canadian to help with ISS construction, first Canadian woman to board the ISS, is now CSA Chief Astronaut	Mission Specialist
Pamela Ann Melroy	US	STS-92 Discovery	11-Oct-00	24-Oct-00	None		Pilot
Peggy Whitson (future)	US	STS 111 Endeavour	May-02			ISS up on STS-111	ISS Crew on STS-111
Laurel Clark (future)	US	STS-107 Columbia	Jun-02				Mission Specialist
Sandra H. Magnus (future)	US	STS-112 Atlantis	Aug-02				Mission Specialist
Heidemarie Stefanyshyn-Piper	US	STS-115 Endeavour	Apr-03				Mission Specialist
Barbara R. Morgan	US	TBD	2004			First Educator Mission Specialist	Educator Mission Specialist

Appendix D:

Timeline of Human Space Exploration

· 1901: Russian Konstantin Tsiolkovsky works out the equations to propel a rocket-powered spacecraft.

· 13 December 1903: The Wright Brothers test fly the first controllable powered heavier than air airplane, the Wright Flyer.

· 16 March 1926: Robert Goddard successfully launches the first liquid fuel rocket.

· September 1944: Germany uses V-2 rockets in attacks against London.

· 1945: World War II ends. A team of top German ballistic missile designers led by Wernher von Braun evades the SS to surrender to the US army. The Russians captured much of von Braun's production team.

· 14 October 1947: Flying the X-1 Glamorous Glennis, Pilot Chuck Yeager becomes the first person to break the sound barrier.

· 4 October 1957: The Soviet Union launches Sputnik I. American President Dwight Eisenhower is not concerned, but the Congress and the public are outraged. Speaker of the House, John McCormick warns of "mass extinction" and fears for the "survival of the free world."

· 3 November 1957: USSR launches the dog Laika into space onboard Sputnik 2. She is the first Earth animal to reach orbit.

· 6 December 1957: Vanguard I fails to launch. In a wave of public humiliation, newspapers called the launch "Kaputnik." Visiting Soviet Leader Nikita Khrushchev watched the launch attempt.

· 31 January 1958: Explorer I launches. First successful US launch of a spacecraft. Discovers Van Allen Radiation Belt around the Earth. Built by the Jet Propulsion Lab.

· 17 March 1958: Vanguard is successfully launched. The first solar powered spacecraft.

· 1 October 1958: the U.S. Congress Space Act forms the National Aeronautics and Space Administration (NASA).

· 1 January 1959: The USSR launches Luna-1, the first spacecraft to escape Earth orbit.

· 7 February 1959: The Lovelace Clinic in Albuquerque, New Mexico begins testing the first group of male *Mercury* astronaut candidates.

· 9 April 1959: First 7 NASA astronauts chosen and announced (Gordon Cooper, Donald K. "Deke" Slayton, John Glenn, Scott Carpenter, Alan Shepard, Virgil "Gus" Grissom, Walter Schirra).

· Fall 1959: Air Force ARDC (Air Research and Defense Command) at Wright Air Development Center conducts astronaut tests on 58-year-old Pilot Ruth Nichols.

· September 1959: Dr. Randolph Lovelace II and Air Force Brig. Gen. Don Flickinger start project WISE (Woman in Space Earliest) at the Air Force ARDC. Lovelace and Flickinger approach Pilot Geraldyn "Jerrie" Cobb as a potential candidate for their astronaut tests.

· 12 September 1959: Soviet spacecraft Luna-2 is the first man made object to reach the Moon.

· 3 October 1959: Soviet spacecraft Luna-3 takes the first photographs of the far side of the Moon.

· 7 December 1959: By this date the WISE program has been cancelled by the Air Force with little explanation (Letter from Flickinger to Cobb made reference to the "unfortunate Nichols release").

· Mid-December 1959: Flickinger turns over the women's testing program to the Lovelace Foundation and Dr. Lovelace.

· February 1960: Cover of Look magazine shows Pilot Betty Skelton in a space suit with the question "Should a Girl Be First in Space?"

· 24 October 1960: An estimated 100 people die when an R-16 ICBM explodes in Baikonur Russia.

· Week of 14 February 1960: Jerrie Cobb undergoes testing at the Lovelace Clinic. Becomes the first woman to pass the same tests as the *Mercury* 7 astronauts.

· 30 January 1961: HAM, (real name Boris) a chimpanzee, makes a sub orbital flight on a *Mercury* Redstone rocket.

· 12 February 1961: The Soviets launch Venera-1 spacecraft towards Venus. Contact with the spacecraft was lost in flight.

· 12 April 1961: USSR Cosmonaut Yuri Gagarin does one orbit of the Earth on *Vostok* I, becomes the first human in space.

· 17 April 1961: Bay of Pigs fiasco. President Kennedy is humiliated.

· 5 May 1961: NASA astronaut Alan Shepard does sub orbital to Bermuda, launched on Redstone rocket on Freedom 7. Shepard meets President John F. Kennedy.

· 25 May 1961: President Kennedy announces goal to send men to the Moon and return to Earth safely. "...Landing a man on the Moon and returning him safely to the Earth ...we choose to go to the Moon in this decade not because it is easy, but because it is hard."

· 21 July 1961: Second American flight, Gus Grissom on Liberty Bell 7 sub-orbital.

· 6 August 1961:USSR Cosmonaut Gherman Titov spends 25 hours in orbit *Vostok* 2.

· 13 August 1961: The Berlin wall is erected.

· 1961: By the end of summer, Dr. Lovelace has tested 25 women.

· September 1961: Dr. Lovelace was to do more testing on 13 women finalists at the naval aviation center in Pensacola Florida. The tests were canceled. The Navy wouldn't support the further testing without an official request from NASA, and NASA refused to sponsor it.

· 11 October 1961: Pilot Bob White in an X-15 sets a new altitude record of 217,000 feet, the edge of space.

· 20 February 1962: NASA Astronaut John Glenn Jr. becomes the first American to orbit the Earth. *Mercury*-Atlas Friendship 7. Orbited 3 times in 5 hours.

· June 1962: Second group of 9 astronauts selected for *Gemini* program out of 253 considered (Neil Armstrong, Frank Borman, Charles "Pete" Conrad, James Lovell, James McDivitt, Elliot See, Thomas Stafford, Edward White, and John Young). A black USAF test pilot, Captain Ed Dwight, is considered but not selected.

· 17-18 July 1962: Subcommittee of the House Committee on Space and Astronautics hearings. Jerrie Cobb and Jane Hart testify for the FLATS. Jacqueline Cochran, John Glenn, Scott Carpenter, and George Low testify the women are not qualified and would hurt space program. John Glenn testified, "The fact that women are not in this field is a fact of our social order."

· October 1962: Cuban Missile Crisis: Probably the closest the United States and the Soviet Union ever came to nuclear war with each other.

· December 1962: US launched Mariner-2 completes the first Venus flyby.

· 15-16 May 1963: Last of the *Mercury* flights, Gordon Cooper orbits for 34 hours 19 minutes in Faith 7 completing 22 orbits.

· 16 June 1963: USSR Cosmonaut Valentina Tereshkova, age 26, becomes first woman in space. She orbits for 2 days, 22 hours, 50 minutes on *Vostok* 6 before returning to Earth on 19 June 1963.

· 28 June 1963 Article by Clare Boothe Luce scolds NASA for not getting a woman into space first: before the USSR.

· October 1963: NASA selects third group of 14 astronauts for *Apollo* program. All selected are white males. Out of the 271 applications, 200 were from civilians including two women who applied, 71 were from military personnel including two black pilots.

· 22 November 1963: President John F. Kennedy is assassinated.

· 12-13 October 1964: The Soviet Union launches the first three-member crew. It is also the first mission with non-pilot astronauts, a physician and an aerospace engineer.

· 1964: Civil Rights Act passed by U.S. Congress.

· 28 June 1964: NASA selects fourth group of Astronauts. It is the first group with non-pilots. 1351 applications are cut down to 400 finalists including 4 women. Only six non-pilots are finally chosen, all men, an engineer, two physicists, two physicians, and a geologist.

· 18 March 1965: USSR Cosmonaut Alexei Leonov makes the first space walk. Voskhod 2.

· 23 March 1965: First American mission with two people on *Gemini* III. First manned flight in the *Gemini* program.

· 3 June 1965: First American EVA by Edward White on *Gemini* IV.

· 15 July 1965: Mariner-4 completes a flyby of Mars.

· 3 February 1966: The Soviet Luna-9 probe conducts the first soft landing and scientific research of the Moon's surface.

· 11-15 November 1966: Last of the *Gemini* flights preparing for *Apollo* program. Successfully conducts a rendezvous station keeping, and docking with Agena. Edwin Aldrin sets an EVA record of 5 hours.

· 27 January 1967: *Apollo* 1 disaster. American astronauts Virgil "Gus" Grissom, Edward White, and Roger Chaffee are killed when fire swept through their *Apollo* 1 capsule in a pre-flight test.

· 24 April 1967: Cosmonaut Vladimir Komarov dies on landing after *Soyuz*-1 test flight.

· 14 July 1968: The explosion of the Proton rocket with L-1 (Zond) spacecraft in pre-launch processing greatly delays the Soviet manned lunar program by destroying the launch complex.

· 11 October 1968: *Apollo* 7. First manned *Apollo* flight (Schirra, Cunningham, and Eisele).

· 21 December 1968: *Apollo* 8: First manned flight around the Moon (Borman, Lovell, Anders).

· 20 July 1969: *Apollo* 11 Moon Landing. First people to set foot on the Moon (Armstrong and Aldrin on the Moon and Collins in orbit).

· 11 April 1970: Launch of *Apollo* 13. At 55 hours into flight "Houston we've had a problem."

· 24 April 1970: China launches its first artificial satellite on a Chinese rocket.

· 1972: *Apollo* 17. Last Moon landing (Cernan and Schmitt).

· 1972: Congress passes amendment to the Civil Rights Act of 1964: the Equal Employment Opportunity Act stating that a federal agency cannot discriminate on basis of race, color, sex, religion or national origin.

· 14 May 1973: The US launches *Skylab*; three crew's visit and work aboard the station.

· 26 December 1974: The Soviet Union launches the Salyut-4 orbital station.

· 17 July 1975: Soviet *Soyuz* and US *Apollo* dock in space.

· 8 July 1976: NASA announcement encourages women and minorities to apply for astronaut program.

· 20 July and 3 September 1976: Viking spacecraft land on Mars and transmit the first images from the surface.

· January 1978: The first class of NASA astronaut candidates for the Shuttle era is chosen. 35 Astronauts including 6 women: Ride, Fisher, Seddon, Resnik, Sullivan, and Lucid. Also the first African-American astronauts (three men) and an Asian man.

· 11 July 1979: *Skylab*'s uncontrolled re-entry from orbit to the Earth.

· 12 April 1981: Launch of *Columbia*, the first Space Shuttle flight (STS-1).

· 19 August 1982: USSR Cosmonaut Svetlana Savitskaya launches into space onboard *Soyuz* T-7. Becomes the second woman to travel into space.

· 18 June 1983: Sally Ride becomes first U.S. woman Astronaut. *Challenger* mission STS-7 deploys communications satellites.

· 3-11February 1984: NASA Astronaut Bruce McCandless becomes the first human satellite when he uses the Manned Maneuvering Unit MMU to perform an un-tethered space walk (shuttle mission 41-B).

· 6-13 April 1984: The first on orbit satellite repair is performed on *Challenger* mission 41-C for the Solar Maximum Mission.

· 1984: The third Soviet expedition to Salyut-7 remains onboard for 237 days, setting a new world record.

· 25 July 1984: USSR Cosmonaut Svetlana Savitskaya becomes the first woman to space walk, *Soyuz* T-12. 3.5 hours of EVA.

· 30 August 1984 Judy Resnik becomes fourth woman to fly in space. Mission 41-D Space Shuttle *Discovery*.

· 5 October 1984: Kathryn Sullivan becomes the first American woman to space walk. Mission 41G. It is the first time 2 women are in space at the same time (Ride and Sullivan).

· 8 November 1984: Anna Fisher, the first mother in space. Mission 51-A deploys two satellites, and retrieves two satellites for repair on Earth, the first time a satellite recovery is ever done.

· 12 April 1985: Margaret Rhea Seddon. Mission 51-D, Communication Satellite deployment, and a repair to correct the propulsion stage of another satellite.

· 17 June 1985: Shannon Lucid. Mission 51-G, Communication Satellites deployed.

· 30 October 1985: Bonnie Dunbar. Mission 61-A a Spacelab mission.

· 26 November 1985: Mary L. Cleave. Mission 61-B demonstrates construction of an erectable structure on orbit, as well as deploying 3 satellites.

· 28 January 1986: Space Shuttle *Challenger* explodes after launch, killing astronauts Dr. Judy Resnik, Commander Francis R. Scobee, pilot Michael.J. Smith (USN), fellow mission specialists, Dr. Ronald. E. McNair, and Lieutenant Colonel Ellison S. Onizuka (USAF), civilian payload specialist Greg. B. Jarvis and the attempt to put the first teacher in space Christa McAuliffe.

· 20 February 1986: The Soviets launch the core module of the *Mir* space station.

· 26 May 1987: Sally Ride announces that she is leaving NASA to begin a two-year fellowship at Stanford's Center for International Security and Arms Control.

· 29 September 1988: Shuttle program restarted with the launch of *Discovery* (STS-26) and deployment of a TDRS satellite.

· 4 May 1989: Mary Cleave's second shuttle flight (STS-30) and the deployment of the Magellan spacecraft to Venus.

· 18 October 1989: Shuttle *Atlantis* (STS-34) with Ellen S. (Shulman) Baker, her first flight, and Shannon Lucid, her second flight, launches the Galileo spacecraft towards Jupiter.

· 22 November 1989: Kathryn C. Thornton is the first woman on a classified US military space flight (STS-33).

· 9 January 1990: Marsha S. Ivins makes her first flight with Bonnie Dunbar, her second flight, on STS-32 which deploys a satellite, and retrieves the Long Duration Exposure Facility (LDEF).

· 24-29 April 1990: Shuttle *Discovery* (STS-31) deploys Hubble Space telescope (STS-31) with mission specialist Kathryn Sullivan's second flight.

· 5 April 1991: Linda M. Godwin's first flight (STS-37) and deployment of the Gamma Ray Observatory.

· 18 May 1991: British chemist/engineer Helen Patricia Sharman flies to the *Mir* space station on a Soviet *Soyuz* TM-12 after she answered a radio advertisement.

· 5 June 1991: Millie Hughes Fulford and Tamara E. Jernigan fly into space for the first time on STS-40 *Columbia*, a Spacelab experimentation mission. Margaret Rhea Seddon flies with them, her second flight, making the first flight with three women on the crew.

· 2 August 1991: Shannon Lucid's third flight on STS-43 and deployment of a TDRS satellite.

· 22 January 1992: Roberta L. Bondar becomes the first Canadian woman in space on STS-42, an International Microgravity Laboratory (IML-1) mission.

· 24 March 1992: Kathryn Sullivan on her third flight is the first female payload commander of *Atlantis* STS-45, Atmospheric Laboratory for Applications and Science (ATLAS-1) mission.

· 7 May 1992: Kathryn Thornton takes her second flight on STS-49. She is part of the first three person EVA.

· 25 June 1992: Bonnie Dunbar's third flight, and Ellen Baker's second on STS-50, US Microgravity Laboratory, and the First Extended Duration Orbiter.

· 31 July 1992: Marsha Ivins takes her second flight on STS-46, a Tethered Satellite System experiment.

· 12 September 1992: First flight of N. Jan Davis. Davis flies with her husband Mark Lee, the first married couple to fly together. Mae Jemison becomes the first African American woman in space. STS-47 *Endeavour* is a Spacelab experiments mission.

· 22 October 1992: Tamara Jernigan takes her second flight on STS-52, a US Microgravity Payload and Deployment of the Laser Geodynamic Satellite.

· 13 January 1993: Major Susan J. Helms (USAF), the first female American military astronaut flies on STS-54, as TDRS satellite deployment.

· 8 April 1993: Ellen Ochoa becomes the first Hispanic woman in space (STS-56) an Atmospheric Laboratory for Applications and Science (ATLAS-2) mission.

· 21 June 1993: Nancy J. (Sherlock) Currie and Janice E. Voss make their first trip into space onboard STS-57 *Endeavour*, a Spacehab-1 mission and retrieves the European Retrievable Carrier (ERC).

· 18 October 1993: Margaret Rhea Seddon makes her third flight, and Shannon Lucid makes her fourth flight on STS-58 as Spacelab life sciences mission.

· 2-13 December 1993 Shuttle crew (STS-61) retrieves, repairs, and redeploys Hubble space telescope with Kathryn Thornton, her second flight.

· 8 January 1994: Russian cosmonaut Valery Polyakov boards *Mir* starting the world's longest spaceflight (18 months).

· 11 Feb 1994: N. Jan Davis second flight (STS-60), as Spacehab-2 mission.

· 4 March 1994: Marsha Ivins third flight (STS-62) a US Microgravity Payload mission.

· 9 April 1994: Linda Godwin second flight (STS-59) a Space Radar Laboratory mission.

· 8 July 1994: Japanese astronaut Chiaki Naito-Mukai, first flight on STS-65 *Columbia* an International Microgravity Laboratory (IML-2) mission.

· 3 November 1994: Ellen Ochoa's second flight (STS-66) with the LIDAR in-space technology experiment, and the Shuttle Pointed Autonomous Research Tool for Astronomy.

· 3 February 1995: Eileen Marie Collins becomes the first woman pilot of a Space Shuttle STS-63 *Discovery* a Spacehab mission. Janice Voss flies with her on her second flight. Several members of the FLATS were at the launch.

· 2 March 1995: Wendy B. Lawrence's first flight on STS-67 Endeavor Ultraviolet Astronomy mission, with Tamara Jernigan, her third flight.

· 27 June 1995: Ellen Baker's third flight, and Bonnie Dunbar's fourth flight (STS-71). US space shuttle docks with Russian *Mir* space station for the first time taking up two Soviet cosmonauts and returning with two others.

· 13 July 1995: Mary Ellen Weber's first flight, and Nancy Currie's second (STS-70) a TDRS deployment mission.

· 20 October 1995: Cathryn G. "Cady" Coleman's first flight, and Kathryn Thornton's fourth on STS-73 *Columbia* a US microgravity laboratory mission.

· 22 March 1996: Shannon Lucid starts her fifth mission to space where she spends 188 days on the *Mir* space station, more than any other American astronaut. Launched on STS-76 with Linda Godwin (her third mission).

· 20 June 1996: Susan Helms third mission (STS-78) a Life and Microgravity Spacelab mission.

· 17 August 1996: Claudie (Andre-DeShays) Haigneré, Russian-French Cassiopée mission to Mir.

· 19 November 1996: Tamara Jernigan's fourth mission (STS-80) orbiting Retrievable Far and Extreme Ultraviolet Spectrometer.

· 22 January 1997: Marsha Ivins takes her fourth flight (STS-81), a Spacehab and *Mir* docking mission.

· 4 April 1997: Pilot Lt Commander Susan L. (Still) Kilrain, USN, takes her first flight, and Janice Voss's third (STS-83) a Microgravity Science Laboratory mission.

· 15 May 1997: Russian Cosmonaut Elena V. Kondakova. STS-84 *Atlantis* a *Mir* docking and Spacehab mission, flown by Eileen Collins, her second flight.

· 25 June 1997: Progress cargo ship collides with *Mir* while attempting to dock, depressurizes the Spektr module.

· 1 July 1997: Susan L. (Still) Kilrain's second flight, Janice Voss's fourth on STS-94 a Microgravity Science Laboratory mission

· 7 August 1997: Jan Davis third flight (STS-85) a Cryogenic Infrared Spectrometer and Telescopes for the Shuttle Pallet Satellite mission.

· 25 September 1997: Wendy Lawrence, second flight (STS-86) a Spacehab and *Mir* docking mission.

· 19 November 1997: Kalpana Chawla first flight on STS-87 a US Microgravity Payload mission.

· 22 January 1998: Bonnie Dunbar fifth flight (STS-89) a Spacehab and *Mir* docking mission.

· 17 April 1998: First flight of Commander Kathryn Hire, USN (STS-90) a Neurolab mission.

· 2 June 1998: Wendy Lawrence third flight, Janet Kavandi first flight (STS-91) a Spacehab and *Mir* docking mission.

· 29 October 1998: Second flight for Japanese astronaut Chiaki Naito-Mukai (STS-95) a Spacehab mission.

· 20 November 1998: The Russians launch the first element of the ISS, the Zarya command module.

· 4-15 December 1998: Space Shuttle (STS-88) launches and docks the Unity module and pressurized mating adaptors 1 and 2 to the Zarya command module to create the first segment of the ISS. Third flight of Nancy Currie.

· 27 May 1999: First flight of Canadian Astronaut Julie Payette with Ellen Ochoa, her third flight, and Tamara Jernigan, her fifth flight, this time with an EVA. STS-96 *Discovery* does an ISS docking and Spacehab mission.

· 23 July 1999: Eileen Collins becomes the first woman to command a Space Shuttle mission. STS-93 *Columbia*, the deployment for the Chandra X-ray Observatory. The second flight of Cathryn Coleman.

· 11 February 2000: Fifth flight of Janice Voss, Second flight of Janet Kavandi. STS-99 is a Shuttle Radar Topography Mission.

· 19 May 2000: First flight of Mary Ellen Weber. Forth for Susan Helms. STS-101 *Atlantis* is an ISS docking and Spacehab mission.

· 11 October 2000: First flight of Pamela Ann Melroy: STS-92 *Discovery* carries an ISS Z-1 truss and pressurized mating adaptor.

· 8 March 2001: Launch of STS-102 *Discovery*. Susan J. Helms spends 163 days aboard the ISS as the first female crewmember. Returned on STS-105.

· 23 March 2001: *Mir* re-enters the Earth's atmosphere and breaks up over the South Pacific.

· April 2001: Space Tourist Dennis Tito visits the ISS.

· 12 July 2001: STS-104 *Atlantis*. ISS assembly flight, the third for Janet Kavandi.

· 5 December 2001: STS-108 *Endeavour*. Linda Godwin's fourth flight and second EVA.

· 2002: all scheduled 2002 Space Shuttle flights have at least 1 woman astronaut onboard.

· 1 March 2002: Nancy (Sherlock) Currie's works the shuttle arm on her fourth flight (STS-109 *Columbia*) to upgrade the Hubble Space Telescope.

· 8 April 2002: STS-110 *Atlantis*. Ellen Ochoa's fourth flight. Truss and mobile transporter integration to the ISS.

· Future planned missions (at the time of this book)
· May 2002: STS-111 *Endeavour*. Peggy Whitson up - ISS crew #5. This will be her 1st flight.

· July 2002: STS-107 *Columbia*. Laurel Clark's first flight. Kalpana Chawla's second flight.

· August 2002: STS-112 *Atlantis*. Sandra H. Magnus's first flight. Shuttle Pilot Pamela Melroy's second flight.

· September 2002: STS-113 *Endeavour*. Peggy Whitson returns with 2 other members of ISS crew #5.

· January 2003: STS-114 *Atlantis*. Eileen Collins fourth flight. Commander of flight.

· April 2003: STS-115 *Endeavour*. Heidemarie Stefanyshyn-Piper's first flight with an EVA.

· 2004: Educator Mission Specialist Barbara Morgan visits the ISS.

Index

About the Author:

Laura S. Woodmansee is a freelance Science Writer. She lives with her rocket scientist husband Paul, and their two dogs Boomer and Goldie. Laura has a Master's in Journalism from the University of Southern California's Annenberg School of Journalism. All her life, Laura has wanted to travel into space.

~EASY TO USE!~ CD-ROM

The attached CD-ROM requires no installation.

It is designed to leave no footprint on your computer's hard-drive and requires no special drivers to run.

All of the files are programmed as a web page and only require a web browser to view. All files are HTML, JPG and MPG.

The disc includes: Exclusive video interviews with Bonnie Dunbar, Heidi Piper, Ellen Ochoa, Ellen Baker, Linda Godwin, Eileen Collins, Kalpana Chawla and Anna Fisher plus much more!